Turfgrass Patch Diseases

Caused by Ectotrophic
Root-Infecting Fungi

Edited by
Bruce B. Clarke and Ann B. Gould
Rutgers—The State University
New Brunswick, New Jersey

APS PRESS
The American Phytopathological Society
St. Paul, Minnesota

Financial Sponsors

BASF Corporation
CIBA-GEIGY Corporation, Agricultural Division
W. A. Cleary Corporation
DowElanco
E. I. du Pont de Nemours and Company
Grace-Sierra Crop Protection Company
LESCO, Inc.
Miles, Inc. (formerly Mobay Corporation)
Monsanto Company
Rhône-Poulenc Ag Company
Rohm and Haas Company
Sandoz Crop Protection Corporation

Cover illustrations (clockwise, from upper left): take-all patch on an *Agrostis* fairway, with the center of the patch recolonized by *Festuca rubra*, courtesy B.B. Clarke; lobed hyphopodia of *Gaeumannomyces graminis* var. *graminis*, reprinted from Compendium of Turfgrass Diseases, 2nd ed., by R.W. Smiley, P.H. Dernoeden, and B.B. Clarke, APS Press, 1992; ascospores and phialospores of *Magnaporthe poae*, reprinted from Compendium of Turfgrass Diseases, 2nd ed.; and perithecia of *G. graminis* var. *avenae* on wheat (*Triticum aestivum*) courtesy P.J. Landschoot.

This publication is based, in part, on presentations from a discussion session entitled "Perspectives on Turfgrass Patch Diseases Caused by Ectotrophic Fungi" held at the annual meeting of The American Phytopathological Society, August 21, 1989, in Richmond, Virginia. This book has been reproduced directly from computer-generated copy submitted in final form to APS Press by the editors of this volume. No editing or proofreading has been done by the Press.

Reference in this publication to a trademark, proprietary product, or company name by personnel of the U.S. Department of Agriculture or anyone else is intended for explicit description only and does not imply approval or recommendation to the exclusion of others that may be suitable.

Library of Congress Catalog Card Number: 93-72095
International Standard Book Number: 0-89054-154-X

© 1993 by The American Phytopathological Society
Second printing, 1994

Printed in the United States of America on acid-free paper

The American Phytopathological Society
3340 Pilot Knob Road
St. Paul, Minnesota 55121-2097, USA

TABLE OF CONTENTS

iii

W. W. Shane, Michigan State University
J. C. Stier and S. T. Nameth, The Ohio State University

P. H. Dernoeden, University of Maryland

Color section follows page 106

PREFACE

Patch diseases that are caused by ectotrophic root-infecting fungi represent a newly recognized and destructive group of turfgrass diseases. Prior to 1984, *Gaeumannomyces graminis* var. *avenae*, the incitant of take-all patch, was the only member of this group recognized as a pathogen of turfgrasses in North America. Since that time, there has been a rapid expansion of knowledge pertaining to the ectotrophic root-infecting fungi. Recent improvements in isolation and detection techniques have contributed to the confirmation of four additional diseases or disease associations. As a result, seven species of ectotrophic fungi, some newly described, have been implicated in patch disease development. This recent surge in activity has made it difficult for both scientists and practitioners to keep abreast of new developments.

In 1989, the committee on Diseases of Ornamentals and Turfgrasses sponsored a discussion session entitled "Perspectives on Turfgrass Patch Diseases Caused by Ectotrophic Fungi" at the annual meeting of the American Phytopathological Society in Richmond, Virginia. The purpose of this session was to summarize the current status of patch disease research and to identify areas that required further investigation. The information presented in this discussion was used as the basis for this book. Since knowledge in this area is rapidly expanding, the authors and editors have made every effort to include the latest information on each topic for this publication. Emphasis has been placed on the biology, taxonomy, epidemiology, and detection of the ectotrophic fungi that cause turfgrass patch diseases. In addition, historical perspectives, symptom development, and management strategies for these patch diseases are presented. It is hoped that researchers, diagnosticians, and practitioners who have an interest in this group of turfgrass pathogens will find this book to be a helpful and stimulating resource.

The publication of this manuscript is the result of a cooperative effort between the authors, reviewers, editors, and financial contributors. We extend very special thanks to Dr. Allison Tally of the CIBA-GEIGY Corporation. In her role as a senior editor for APS Press, her guidance, suggestions, and encouragement during the editorial process were very much appreciated. We also wish to acknowledge Dr. Gary Chastagner, Washington State University, and Dr. Ned Tisserat, Kansas State University, for their insightful and thorough reviews of the manuscript. To Dr. Gail Schumann, University of Massachusetts, we thank you for assembling many of the color plates published in this volume. We are also very grateful to Dr. Donald Kobayashi, Dr. Bradley Hillman, Dr. Marshall Bergen, and Mr. Pradip Majumdar for their advice and technical assistance.

Finally, we would like to recognize the following Universities for allowing the authors to contribute their time and materials throughout the preparation of this publication: The University of Maryland System, Michigan State University, The Ohio State University, Oregon State University, The Pennsylvania State University, University of Rhode Island, and Rutgers, The State University of New Jersey.

Bruce B. Clarke and Ann B. Gould
Department of Plant Pathology
Rutgers, The State University
New Brunswick, NJ 08903

HISTORICAL PERSPECTIVE OF RESEARCH ON ECTOTROPHIC ROOT-INFECTING PATHOGENS OF TURFGRASSES

Richard W. Smiley
Oregon State University
Columbia Basin Agricultural Research Center
Pendleton, Oregon 97801

INTRODUCTION

Root-infecting fungi that cause diseases of turfgrasses have become the focus of intensive research in North America. In particular, attention has turned to those pathogens that utilize an ectotrophic growth habit to colonize roots before penetrating vascular tissue. The rapid expansion of knowledge pertaining to this group of fungi has made it difficult for scientists, as well as diagnosticians and practitioners, to keep abreast of new developments. This becomes particularly apparent when one considers that prior to 1984, *Gaeumannomyces graminis* (Sacc.) Arx & D. Olivier var. *avenae* (E. M. Turner) Dennis was the only member of this group recognized as a pathogen of turfgrasses in North America. The list of ectotrophic fungi now includes seven confirmed turfgrass pathogens representing four distinct genera.

This chapter presents an historical perspective of past research and observations that have led to our current understanding of ectotrophic root-infecting fungi in turfgrasses. From this work, five diseases or disease associations have been reported. As our knowledge in this area increases, it is likely that new host-pathogen associations will be identified.

TAKE-ALL PATCH

Ophiobolus graminis (Sacc.) Sacc. in Roum. & Sacc. was independently reported to be associated with a patch disease of turfgrass in Holland (30) and in the United States (25) during 1931 and 1932, respectively. By the time *O. graminis* was first isolated from diseased turf, it was already a recognized pathogen of cereals. As a result, little uncertainty existed regarding the causal nature of this patch disease (28). In 1956, the disease was named Ophiobolus patch (40). *O. graminis* has subsequently been associated with patch diseases of turfgrass in other European countries and in Australia (40).

Although *O. graminis* was first implicated as a pathogen of turfgrass in North America during the early 1930s, the identification and pathogenicity of this fungus was not actually confirmed until 1960 in western Washington (13) and 1978 in Massachusetts and Rhode Island (14, 16). Difficulties in isolating the pathogen from older, infected tissues (6) apparently resulted in the 30 to 50 year delay in confirming the causal agent and in specifying the distribution of this disease in North America. By the mid-1980s, it had become clear that the disease occurs throughout North America (Fig. 2.1).

Refinements in the taxonomy of *Ophiobolus* have led to several changes in the name of the causal agent of Ophiobolus patch. In 1952, several species of *Ophiobolus* were transferred into the newly erected genus *Gaeumannomyces* Arx & D. Olivier (45). As a result, *O. graminis* was renamed *G. graminis* (Sacc.) Arx & D. Olivier. Two decades later, *G. graminis* was divided into three distinct varieties (*G. graminis* (Sacc.) Arx & D. Olivier var. *graminis*; *G. graminis* var. *avenae*; and *G. graminis* (Sacc.) Arx & D. Olivier var. *tritici* J. Walker) (48). The taxonomic status of this group has been extensively reviewed by Walker (49), and a guide to diagnostic procedures has been developed by Smiley et al (38). The name

2

of the disease was later changed to Gaeumannomyces patch in 1980 (15) and finally to take-all patch in 1981 (16). Take-all patch is currently the preferred name for use in publications sponsored by the American Phytopathological Society (3).

G. *graminis* var. *avenae* is a pathogen of a number of grass hosts; however, bentgrasses (*Agrostis* spp.) are especially susceptible to infection (41). Although *G. graminis* var. *avenae* is the recognized incitant of take-all patch, *G. graminis* var. *tritici* has also been associated with this disease in Australia (52). This should not come as a surprise in view of the high phenotypic variability among isolates of these varieties (27, 54, 55). *G. graminis* var. *avenae* and *G. graminis* var. *tritici* are separated primarily on the basis of ascospore length and pathogenicity to oat (*Avena* spp.). Intermediate forms of these fungi are common because both characteristics are extremely variable. In fact, ascospore length may occur in a continuum from one variety to the other (50).

The genus *Gaeumannomyces* has been more intensively studied than any other root- or crown-infecting pathogen of the Gramineae. This wealth of information, collected primarily from research involving take-all of small grains, has been an important resource for scientists studying take-all and related patch diseases of turf (1). The development of a semi-selective culture medium in 1984 to isolate *Gaeumannomyces* from roots (17) further improved the ability of diagnosticians and researchers to work with this group of fungi.

SPRING DEAD SPOT

Spring dead spot, a serious patch disease of bermudagrass (*Cynodon dactylon* (L.) Pers.), was first observed in Oklahoma during 1936 (41). The disease was not officially named spring dead spot, however, until 1960 (47). The identity of the primary agent(s) has become the subject of a spirited debate that continues today, more than 50 years after the disease was first described. In Australia, a disease of similar etiology was first noted on *Cynodon* spp. in 1961 (39). This disease also became known as spring dead spot. The first advance in determining the causality of spring dead spot in eastern Australia came from studies conducted by Smith (39). Smith proved that the disease in New South Wales was caused by a soilborne fungus that he initially designated as *Ophiobolus herpotrichus* (Fr.:Fr.) Sacc. & Roum. It was later determined, however, that two distinct fungi were combined within this nomenclatural assignment and that neither of them could be retained within the genus *Ophiobolus*. In 1972, these fungi were described as *Leptosphaeria narmari* J. C. Walker & A. M. Sm. and *L. korrae* J. C. Walker & A. M. Sm. (51), two new species within the genus *Leptosphaeria* Ces. & De Not., nom. cons. *L. narmari* is currently the dominant pathogen associated with spring dead spot in Australia (51).

Helminthosporium spiciferum (Bainier) J. Nicot (= *Bipolaris spicifera* (Bainier) Subramanian) was one of the first agents associated with spring dead spot in the United States (46). Although isolates of this fungus caused leafspots on bermudagrass seedlings, symptoms of spring dead spot were never reproduced (22). Two decades after Smith made his initial report, scientists in the United States began to develop lines of evidence to support the concept that *Gaeumannomyces*-like ectotrophs incite spring dead spot in North America. Agents associated with the disease complex in the United States are now reported to include *L. korrae* (5, 10), *Ophiosphaerella herpotricha* (Fr.:Fr.) J. C. Walker (43, 44), and *G. graminis* var. *graminis* (24).

O. herpotricha is closely related to *L. korrae* (49, 50). Synonyms of *O. herpotricha* include *O. herpotrichus* and *Phaeosphaeria herpotricha* (Fr.:Fr.) L. Holm. Both *O. herpotricha* and *L. korrae* attack roots, stolons, and rhizomes of grasses. *G. graminis* var. *graminis* and its purported anamorph, *Phialophora graminicola* (Deacon) J. Walker, have also been found in association with grasses and cereals (7). Prior to 1988, *G. graminis* var. *graminis* and *O. herpotricha* were generally considered to be saprophytes (7, 50). These fungi, however, are now recognized as additional members of

4

the ectotrophic ascomycetes that cause serious diseases of turfgrasses (8, 24, 43, 44).

In the past, failure to consistently isolate and identify *Gaeumannomyces*-like fungi from turfgrasses exhibiting symptoms of spring dead spot contributed to the confusion surrounding this disease. Some researchers still believe that fungi readily isolated from symptomatic areas, such as species of *Bipolaris*, *Curvularia*, *Drechslera*, or *Fusarium*, are also responsible for spring dead spot. To date, however, scientists have been unable to reproduce typical symptoms of spring dead spot by inoculating turfgrasses with any of these fungi.

Extensive research has been conducted over the past fifty years to develop effective control strategies for spring dead spot. An introduction to this literature is presented by Smith et al (41), Lucas (22), and Dernoeden (chapter six). Unfortunately, the etiology of this disease was unknown when most of this work was conducted. Recent reports detailing the ecological characteristics of the pathogens and the etiology of spring dead spot will undoubtedly serve as an aid in the development of more precise control strategies.

NECROTIC RING SPOT

The etiology of necrotic ring spot has followed a developmental history parallel to that of spring dead spot (36). Necrotic ring spot was first observed on Kentucky bluegrass (*Poa pratensis* L.) in the northeastern United States during the 1950s and was subsequently named Fusarium blight in 1966 (4). The disease, however, could not be reproduced by inoculating turfgrasses with the purported incitants [*F. culmorum* (Wm. G. Sm.) Sacc. (reported as *F. roseum* (LK.) Snyd. & Hans. f. sp. *cerealis* (Cke.) Snyd. & Hans. 'Culmorum') and *F. poae* (Peck) Wollenweb. (reported as *F. tricinctum* (Cda.) Snyd. & Hans. f. sp. *poae* (Pk.) Snyd. & Hans.)]. Moreover, disease expression and control strategies appeared to vary from one geographic region to another (34, 36). As a result, several faculty positions were established at universities in the United States in response to industry concerns about the inability of scientists to understand and control Fusarium blight. This issue became even more controversial when a basidiomycete-like fungus, *Rhizoctonia cerealis* Van der Hoeven, and an unidentified *Gaeumannomyces*-like fungus were suggested as possible incitants of patch diseases similar to those associated with Fusarium blight. These reports were summarized by Smiley (36), and serve to illustrate the diagnostic difficulties faced by scientists working on the frontier of turfgrass pathology.

After two decades of research, it became apparent that a composite of diseases had been grouped together under the name Fusarium blight (36). Components of this group are still in the process of being separated into specific diseases. Necrotic ring spot and summer patch have recently emerged as two such segregants. Like spring dead spot, much of the research on disease control was conducted before the true etiology of Fusarium blight and its segregants was established (29, 34). Concurrent research by turf pathologists at New York, Rhode Island, Washington, and Wisconsin during the early 1980s identified *L. korrae* as the incitant of necrotic ring spot (37, 53). Much of this work, however, has not yet been published in the scientific literature.

We now realize that necrotic ring spot causes serious damage to Kentucky bluegrass turfs in most locations in North America where the grass is grown (41, 53). Extensive efforts are currently in progress to fully describe the biology of *L. korrae* and to develop control strategies for this disease. Studies have been conducted to determine the effects of temperature, moisture, and other environmental parameters on the growth of the pathogen. Each of the major turfgrass species have also been evaluated for seedling resistance to *L. korrae* (36, 53). Efforts to develop monoclonal antibodies (26, 32) and DNA probes (42) specific to *L. korrae* should greatly improve our ability to rapidly and accurately diagnose this disease in infested turfs.

SUMMER PATCH

Summer patch is another recent segregant from the group of patch diseases that affect *Poa* spp. in North America (36). The disease remained an unidentified component of Fusarium blight (Plate 24) until it became apparent that infection by an ectotrophic fungus preceded colonization by *Fusarium* spp. and other secondary fungi (37). In 1984, the ectotrophic agent was described by its conidial state (35) since the teleomorph could not be induced at that time. The fungus conformed closely to the original description of *P. graminicola* (49) and was, therefore, assigned that designation by Smiley and Craven Fowler (37).

A close examination of Smiley's *P. graminicola* isolates by Landschoot and Jackson (21) led to the determination that the collection included a composite of several different fungi. Landschoot and Jackson found that the least virulent and avirulent isolates were indeed *P. graminicola*. The moderately and highly virulent isolates, however, were determined to be a different *Phialophora* species. This species was later confirmed to be the anamorph of *Magnaporthe poae* Landschoot & Jackson (21), the causal agent of summer patch. This represented a new species in the genus *Magnaporthe* R. Krause & R. E. Webster and was the first account of a heterothallic ectotroph parasitizing turfgrass roots. A second heterothallic fungus with a *Phialophora* anamorph was later isolated from the roots of several turfgrass species. This newly identified fungus, named *G. incrustans* Landschoot & Jackson (20), was the first known heterothallic species in the genus *Gaeumannomyces*. In 1990, Kemp et al (18) confirmed the pathogenicity of this fungus on fine-leaved fescues (*Festuca* spp.) under field conditions. Landschoot and Jackson's discovery of heterothallic members in this group of pathogens provides new hope for the taxonomic placement of *Phialophora*-induced turfgrass diseases.

In related findings, Scott and Deacon (31) identified *M. rhizophila* Scott & Deacon as a pathogen of wheat roots in

South Africa. *M. rhizophila* is a homothallic fungus that produces a *Phialophora* stage on plants and in culture that is virtually indistinguishable from that produced by *G. graminis* var. *tritici* and *G. graminis* var. *avenae*. Another isolate, which appeared to be nearly identical on the basis of anamorphic characteristics, failed to produce a teleomorph in the single-isolate tests (31). It is possible that the latter isolate may also be an anamorph of a heterothallic species. Efforts continue to fully describe the biology of *M. poae* and to develop control strategies for this disease. Each of the major turfgrass species have also been evaluated for resistance to *M. poae* (19, 36).

BERMUDAGRASS DECLINE

In 1984, Freeman and Augustin described the first occurrence of a decline that affects closely-mowed bermudagrass turf in Florida (11). A fungus possessing brown, ectotrophic runner hyphae was consistently observed on the surface of affected roots. The authors postulated that this ectotrophic fungus belonged to the genus *Gaeumannomyces* or *Leptosphaeria*. In 1986, one isolate was identified as *Phialophora radicicola* Cain var. *graminicola* Deacon (= *P. graminicola*) (12), a reported anamorph of *G. graminis* var. *graminis*. At the time, this fungus was primarily considered a pathogen of corn (2) and wheat (33).

In 1987 and 1988, Elliott and Landschoot (9) isolated 17 *Gaeumannomyces*-like fungi from the roots of declining bermudagrass turf at nine field locations in Florida. These fungi were identified as *G. graminis* var. *graminis*, *G. incrustans*, *Phialophora* spp., and a sterile *Gaeumannomyces* isolate. At several sites, more than one of these fungi were associated with symptomatic plant roots. *In vitro* pathogenicity studies showed that *G. graminis* var. *graminis* and *G. incrustans* were pathogenic to wheat (8, 9), but only *G. graminis* var. *graminis* was pathogenic to bermudagrass under the conditions tested (8). From this work, *G. graminis* var. *graminis* was determined to be the primary incitant of

8

bermudagrass decline (8); however, confirmation of Koch's postulates for this fungus under field conditions has yet to be established. *G. graminis* var. *graminis* has also been associated with spring dead spot of bermudagrass in North Carolina (23, 24).

Although Elliott (8) found that *G. incrustans* was not pathogenic to bermudagrass *in vitro*, its pathogenicity on wheat suggests that it may contribute to a decline of bermudagrass caused by a complex of *Gaeumannomyces* species (9). Two isolates of *M. poae*, one isolated from bentgrass turf in Florida and the other from Kentucky bluegrass in New York (ATCC 64413), were also pathogenic to wheat and bermudagrass *in vitro*. Although *M. poae* has not been isolated from bermudagrass, its pathogenicity to bermudagrass *in vitro* suggests that *M. poae* may also contribute to bermudagrass decline in the field (8). The role of other fungi associated with symptomatic roots (i.e., *Phialophora* spp. and the sterile *Gaeumannomyces* isolates) remains poorly understood.

SUMMARY

Advances in our understanding of ectotrophic soilborne pathogens that cause patch diseases of turfgrasses are presented. The identification of new members in the complex of fungi that cause these diseases has dominated the progress made in turfgrass pathology during the past decade. At least five new patch diseases and their causal agents have been reported in North America since 1984. *L. korrae* and *M. poae* were identified as incitants of necrotic ring spot and summer patch, respectively. *L. korrae* and *O. herpotricha* became recognized as incitants of spring dead spot. *G. graminis* var. *graminis* was associated with spring dead spot and bermudagrass decline, and *G. incrustans* was described as a root-infecting species in turf. It also became apparent that take-all patch (*G. graminis* var. *avenae*) is more widespread than previously reported. This progress has contributed significantly to the ability of scientists and private industry to develop effective disease identification and control strategies. Studies on the ecology, taxonomy, and physiology of these newly recognized pathogens continue. A detailed account of the etiology and control of these pathogens and diseases is presented in the following chapters of this book.

LITERATURE CITED

1. Asher, M. J. C., and Shipton, P. J., eds. 1981. Biology and Control of Take-all. Academic Press, New York.
2. Cain, R. F. 1952. Studies of Fungi Imperfecti. I. *Phialophora.* Can. J. Bot. 30:338-343.
3. Couch, H. B. 1985. Common names for turfgrass diseases. Plant Dis. 69:672-675.
4. Couch, H. B., and Bedford, E. R. 1966. Fusarium blight of turfgrasses. Phytopathology 56:781-786.
5. Crahay, J. N., Dernoeden, P. H., and O'Neill, N. R. 1988. Growth and pathogenicity of *Leptosphaeria korrae* in bermudagrass. Plant Dis. 72:945-949.
6. Cunningham, P. C. 1981. Isolation and culture. Pages 103-123 in: Biology and Control of Take-all. M. J. C. Asher and P. J. Shipton, eds. Academic Press, New York.
7. Deacon, J. W. 1981. Ecological relationships with other fungi: Competitors and hyperparasites. Pages 75-101 in: Biology and Control of Take-all. M. J. C. Asher and P. J. Shipton, eds. Academic Press, New York.
8. Elliott, M. L. 1991. Determination of an etiological agent of bermudagrass decline. Phytopathology 81:1380-1384.
9. Elliott, M. L., and Landschoot, P. J. 1991. Fungi similar to *Gaeumannomyces* associated with root rot of turfgrasses in Florida. Plant Dis. 75:238-241.
10. Endo, R. M., Ohr, H. D., and Krausman, E. M. 1985. *Leptosphaeria korrae*, a cause of the spring dead spot disease of bermudagrass in California. Plant Dis. 69:235-237.

11. Freeman, T. E., and Augustin, B. J. 1984. Bermuda-grass decline. Fact Sheet PP-31. University of Florida, Gainesville.

12. Freeman, T. E., and Augustin, B. J. 1986. Association of *Phialophora radicicola* Cain with declining bermudagrass in Florida. (Abstr.) Phytopathology 76:1057.

13. Gould, C. J., Goss, R. L., and Eglitis, M. 1961. Ophiobolus patch disease of turf in western Washington. Plant Dis. Rep. 45:296-297.

14. Jackson, N. 1979. More turf diseases; old dogs and new tricks. J. Sports Turf Res. Inst. 55:163-166.

15. Jackson, N. 1980. Gaeumannomyces (Ophiobolus) patch disease. Pages 141-143 in: Advances in Turfgrass Pathology. P. O. Larsen and B. G. Joyner, eds. Harcourt Brace Jovanovich, Duluth, MN.

16. Jackson, N. 1981. Take-all patch (Ophiobolus patch) of turfgrasses in the northeastern United States. Pages 421-424 in: Proc. Int. Turfgrass Res. Conf., 4th. R. W. Sheard, ed. Ontario Agricultural College, University of Guelph and the International Turfgrass Society, Guelph, Ontario.

17. Juhnke, M. E., Mathre, D. E., and Sands, D. C. 1984. A selective medium for *Gaeumannomyces graminis* var. *tritici*. Plant Dis. 68:233-236.

18. Kemp, M. L., Clarke, B. B., and Funk, C. R. 1990. The susceptibility of fine fescues to isolates of *Magnaporthe poae* and *Gaeumannomyces incrustans*. (Abstr.) Phytopathology 80:978.

19. Kemp, M. L., Landschoot, P. J., Clarke, B. B., and Funk, C. R. 1990. Response of fine fescues to field inoculation with summer patch. Page 176 in: Agronomy Abstracts. American Society of Agronomy, Madison, WI.

20. Landschoot, P. J., and Jackson, N. 1989. *Gaeumanno-myces incrustans* sp. nov., a root-infecting hyphopodiate

fungus from grass roots in the United States. Mycol. Res. 93:55-58.

21. Landschoot, P. J., and Jackson, N. 1989. *Magnaporthe poae* sp. nov., a hyphopodiate fungus with a *Phialophora* anamorph from grass roots in the United States. Mycol. Res. 93:59-62.

22. Lucas, L. T. 1980. Spring deadspot of bermudagrass. Pages 183-187 in: Advances in Turfgrass Pathology. P. O. Larsen and B. G. Joyner, eds. Harcourt Brace Jovanovich, Duluth, MN.

23. McCarty, L. B., and Lucas, L. T. 1988. Identification and suppression of spring dead spot disease in bermudagrass. Page 154 in: Agronomy Abstracts. American Society of Agronomy, Madison, WI.

24. McCarty, L. B., and Lucas, L. T. 1989. *Gaeumanno-myces graminis* associated with spring dead spot of bermudagrass in the southeastern United States. Plant Dis. 73:659-661.

25. Monteith, J., Jr., and Dahl, A. S. 1932. Turf diseases and their control. Bull. U. S. Golf Assoc. Green Sect. 12:156-157.

26. Nameth, S. T., Shane, W. W., and Stier, J. C. 1990. Development of a monoclonal antibody for detection of *Leptosphaeria korrae*, the causal agent of necrotic ringspot disease of turfgrass. Phytopathology 80:1208-1211.

27. Nilsson, H. E. 1972. The occurrence of lobed hyphopodia on an isolate of the take-all fungus, "*Ophiobolus graminis* Sacc.," on wheat in Sweden. Swedish J. Agric. Res. 2:105-118.

28. Nilsson, H. E., and Smith, J. D. 1981. Take-all of grasses. Pages 433-448 in: Biology and Control of Take-all. M. J. C. Asher and P. J. Shipton, eds. Academic Press, New York.

29. Sanders, P. L., and Cole, H., Jr. 1981. The Fusarium diseases of turfgrass. Pages 195-209 in: *Fusarium*: Diseases, Biology and Taxonomy. P. E. Nelson, T. A.

Toussoun, and R. J. Cook, eds. Pennsylvania State University Press, University Park, PA.

30. Schoevers, T. A. C. 1937. Some observations on turf-diseases in Holland. J. Board Greenskeep. Res. 5:23-26.

31. Scott, D. B., and Deacon, J. W. 1983. *Magnaporthe rhizophila* sp. nov., a dark mycelial fungus with a *Phialophora* conidial state, from cereal roots in South Africa. Trans. Br. Mycol. Soc. 81:77-81.

32. Shane, W. W., and Nameth, S. T. 1988. Monoclonal antibodies for diagnosis of necrotic ring spot of turfgrass. (Abstr.) Phytopathology 78:1521.

33. Sivasithamparam, K. 1975. *Phialophora* and *Phialophora*-like fungi occurring in the root region of wheat. Aust. J. Bot. 23:193-212.

34. Smiley, R. W. 1980. Fusarium blight of Kentucky bluegrass: New perspectives. Pages 155-175 in: Advances in Turfgrass Pathology. P. O. Larsen and B. G. Joyner, eds. Harcourt Brace Jovanovich, Duluth, MN.

35. Smiley, R. W. 1984. "Fusarium blight syndrome" re-described as a group of patch diseases caused by *Phialophora graminicola*, *Leptosphaeria korrae*, or related species. (Abstr.) Phytopathology 74:811.

36. Smiley, R. W. 1987. The etiologic dilemma concerning patch diseases of bluegrass turfs. Plant Dis. 71:774-781.

37. Smiley, R. W., and Craven Fowler, M. 1984. *Leptosphaeria korrae* and *Phialophora graminicola* associated with Fusarium blight syndrome of *Poa pratensis* in New York. Plant Dis. 68:440-442.

38. Smiley, R. W., Kane, R. T., and Craven-Fowler, M. 1985. Identification of *Gaeumannomyces*-like fungi associated with patch diseases of turfgrasses in North America. Pages 609-618 in: Proc. Int. Turfgrass Res. Conf., 5th. F. Lemaire, ed. INRA Publications and the International Turfgrass Society, Versailles, France.

39. Smith, A. M. 1965. *Ophiobolus herpotrichus*, a cause of spring dead spot in couch turf. Agric. Gaz. N. S. W. 76:753-758.

40. Smith, J. D. 1956. Fungi and turf diseases. (6) Ophiobolus patch disease. J. Sports Turf Res. Inst. 9:180-202.

41. Smith, J. D., Jackson, N., and Woolhouse, A. R. 1989. Fungal Diseases of Amenity Turf Grasses. E. & F. N. Spon, London.

42. Tisserat, N. A., Hulbert, S. H., and Nus, A. 1991. Identification of *Leptosphaeria korrae* by cloned DNA probes. Phytopathology 81:917-921.

43. Tisserat, N., Pair, J., and Nus, A. 1988. *Ophiosphaerella herpotricha* associated with spring dead spot of bermudagrass in Kansas. (Abstr.) Phytopathology 78:1613.

44. Tisserat, N. A., Pair, J. C., and Nus, A. 1989. *Ophiosphaerella herpotricha*, a cause of spring dead spot of bermudagrass in Kansas. Plant Dis. 73:933-937.

45. von Arx, J. A., and Olivier, D. L. 1952. The taxonomy of *Ophiobolus graminis* Sacc. Trans. Br. Mycol. Soc. 35:29-33.

46. Wadsworth, D. F., Houston, B. R., and Peterson, L. J. 1968. *Helminthosporium spiciferum*, a pathogen associated with spring dead spot of bermuda grass. Phytopathology 58:1658-1660.

47. Wadsworth, D. F., and Young, H. C., Jr. 1960. Spring dead spot of bermudagrass. Plant Dis. Rep. 44:516-518.

48. Walker, J. 1972. Type studies on *Gaeumannomyces graminis* and related fungi. Trans. Br. Mycol. Soc. 58:427-457.

49. Walker, J. 1980. *Gaeumannomyces, Linocarpon, Ophiobolus* and several other genera of scolecospored ascomycetes and *Phialophora* conidial states, with a note on hyphopodia. Mycotaxon 11:1-129.

50. Walker, J. 1981. Taxonomy of take-all fungi and related genera and species. Pages 15-74 in: Biology and Control of Take-all. M. J. C. Asher and P. J. Shipton, eds. Academic Press, New York.

51. Walker, J., and Smith, A. M. 1972. *Leptosphaeria narmari* and *L. korrae* spp. nov., two long-spored pathogens of grasses in Australia. Trans. Br. Mycol. Soc. 58:459-466.

52. Wong, P. T. W., and Siviour, T. R. 1979. Control of Ophiobolus patch in *Agrostis* turf using avirulent fungi and take-all suppressive soils in pot experiments. Ann. Appl. Biol. 92:191-197.

53. Worf, G. L., Stewart, J. S., and Avenius, R. C. 1986. Necrotic ring spot disease of turfgrass in Wisconsin. Plant Dis. 70:453-458.

54. Yeates, J. S. 1986. Ascospore length of Australian isolates of *Gaeumannomyces graminis*. Trans. Br. Mycol. Soc. 86:131-136.

55. Yeates, J. S., and Parker, C. A. 1986. In vitro reaction of Australian isolates of *Gaeumannomyces graminis* to crude oat extracts. Trans. Br. Mycol. Soc. 86:137-144.

GEOGRAPHIC DISTRIBUTION, HOST RANGE, AND SYMPTOMATOLOGY OF PATCH DISEASES CAUSED BY SOILBORNE ECTOTROPHIC FUNGI

Noel Jackson
University of Rhode Island
Kingston, RI 02881

INTRODUCTION

Various soilborne fungi attack the roots and crowns of turfgrasses resulting in disease symptoms where grass plants die in patches that are roughly circular, ring, or frogeye in shape. When activity of the causal fungus is restricted to a specific host under a particular set of environmental conditions, then diagnosis using visible symptoms can be attempted with a fair degree of accuracy. If the pathogen produces distinctive signs, such as fruiting structures that are readily discernable, then confirmation of the diagnosis is assured.

In many situations, typical patch symptoms may occur, but the only signs on dead plants are assorted fungal mycelia that are barely visible with a hand lens. Infection and critical damage caused by a primary pathogen can also substantially predate symptom expression so that, when symptoms do occur, evidence of the primary incitant is often compromised by an abundant overgrowth of secondary fungi and bacteria. The only recourse in such cases is to submit samples for laboratory examination. In the laboratory, patch disease fungi may be isolated directly from recently infected tissues or indirectly using bait species (e.g., wheat seedlings) planted into infested soil or plant debris. Although these procedures are time consuming, they are occasionally required for any of the five diseases described below.

TAKE-ALL PATCH

Producers of small grains throughout the world are familiar with the depredations of the common soilborne fungus *Gaeumannomyces graminis* (Sacc.) Arx & D. Olivier. Formerly known as *Ophiobolus graminis* (Sacc.) Sacc. in Roum & Sacc. (41), this fungus is the cause of a root rot disease called take-all or whiteheads that drastically reduces grain yields of cereals (28). Three varieties are recognized within the species *G. graminis*. Each is capable of attacking, with varying severity, cereals and other members of the *Poaceae* (43). One of the trio, *G. graminis* (Sacc.) Arx & D. Olivier var. *avenae* (E. M. Turner) Dennis, is an aggressive pathogen of bentgrasses (*Agrostis* spp.). The resulting disease, designated as take-all patch, severely damages bentgrass turf in Western Europe, North America (Fig. 2.1), and Australasia (Australia and New Zealand) (4, 39).

Host Range

Nilsson (27) and Nilsson and Smith (28) provide extensive literature reviews pertaining to the host range and susceptibility of grass species to *G. graminis*. *G. graminis* var. *avenae* can infect a number of cool-season grasses that may occur as intended or volunteer components of fine turf. Although take-all patch has been reported on annual bluegrass (*Poa annua* L.), rough bluegrass (*P. trivialis* L.), Kentucky bluegrass (*P. pratensis* L.), and velvetgrass (*Holcus lanatus* L.), severe symptoms are usually confined to turfs of colonial bentgrass (*Agrostis tenuis* Sibth.), creeping bentgrass (*A. stolonifera* L.), and velvet bentgrass (*A. canina* L.) (38, 39).

18

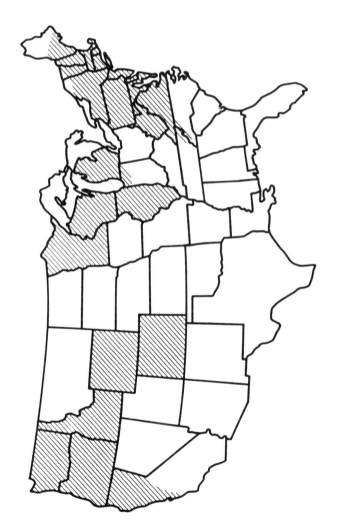

Figure 2.1.   Geographical distribution of take-all patch in the United States.

Symptoms

G. graminis var. *avenae* is active during cool, moist weather. Symptoms of take-all patch may develop in late spring, but are most noticeable during late summer and early fall (15, 28). Bentgrass turf affected by take-all patch dies in roughly circular patches that may enlarge from a few cm to 1 m or more in diameter (Plates 49-53) during the course of several growing seasons (14). Adjacent patches may coalesce, resulting in large, irregularly shaped areas of damaged turf (Plates 54-57). Symptoms are often intensified under conditions of heat or drought stress (Plate 57). In heavily thatched turf, patches form shallow depressions that may impair playing surfaces on sports turf (38, 39).

At the advancing margins of patches, infected plants become yellow or reddish bronze and later turn tan or brown as they die (Plates 54 and 55). Severely infected plants have little root hold and detach easily as the fungus progressively invades host tissue (9). A critical examination of infected plants reveals brown, ectotrophic mycelium on roots, rhizomes, and stolons. Single- or multiple-stranded runner hyphae investing these organs form simple hyphopodia (Plate 60) or complex aggregates of plate mycelium from which colorless infection hyphae are generated. Although infected epidermal and cortical root cells may appear healthy, adjacent stelar tissues are discolored. Vascular tissues in the roots and crown become plugged, turn brown, and die. Extensive mats of dark-colored plate mycelia occur on and between the basal leaf sheaths (28, 38, 39).

Field Diagnosis

Take-all patch is an increasing problem of bentgrass turfs growing under conditions of cool temperatures, ample surface moisture, and high soil or rhizosphere pH (15, 39). The appearance of yellow patches with bronzed margins is indicative of take-all patch in the field. The bases of infected plants are discolored and invested with dark-colored mycelia (Plate 59). These plants detach easily from the soil and can be examined with a hand lens for the presence of black, flask-shaped perithecia (Plates 61 and 62) that develop readily on necrotic leaf sheath tissues in late fall and early spring (41, 43).

In mixed swards, susceptible bentgrasses are killed, leaving behind the other grass components (Plate 53). Disease patches occurring in pure stands of bentgrass are typically colonized by volunteer species such as annual bluegrass and broad-leaved weeds. Bentgrass established on recently cleared woodland sites is particularly prone to infection. New plantings of bentgrass on soils high in sand content are also susceptible to take-all, especially following fumigation (28, 35). Presumably, this relates to the paucity of microbial antagonists in these situations.

Symptoms of take-all patch may persist over a number of years with varied intensity. If left untreated, the disease eventually subsides in a manner similar to that of take-all in cereals (27, 28) (Plate 58). This phenomenon, known as take-all decline, has been attributed to the gradual build-up of soil microbial populations that are suppressive to the take-all pathogen (32).

SUMMER PATCH

Summer patch now constitutes one of the more important problems confronting turf managers in North America (8). First named in 1984, this destructive disease of *Poa* and fine-leaved *Festuca* species has a much longer history as an unidentified component of the Fusarium blight syndrome (33). In 1984, Smiley and Craven Fowler (34) reported that a *Phialophora* species was primarily associated with the occurrence of Fusarium blight on Kentucky bluegrass turf in New York state. This species was later proven to be the anamorph of *Magnaporthe poae* Landschoot & Jackson (23), the causal agent of summer patch. To date, summer patch has only been reported in North America (Fig. 2.2).

21

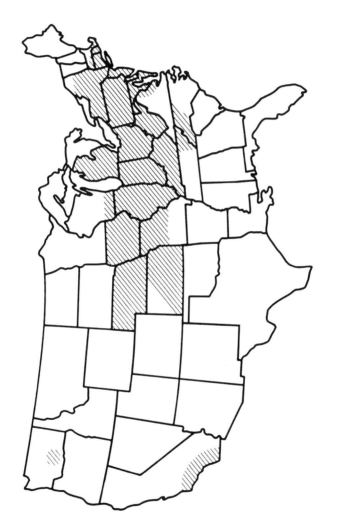

Figure 2.2.   Geographical distribution of summer patch in the United States.

Host Range

The pathogenicity of *M. poae* has been confirmed on Kentucky bluegrass (23), annual bluegrass (23), and fine-leaved fescues (19, 20, 21). Although this disease is most often associated with Kentucky bluegrass turf, annual bluegrass, a common volunteer constituent of greens and fairways, is also highly susceptible to *M. poae*. As a result, summer patch is now recognized as a widespread and destructive problem of both annual and Kentucky bluegrass turf (8). *M. poae* has also been isolated on occasion from creeping bentgrass and from perennial ryegrass (*Lolium perenne* L.) (35). Creeping bentgrass, perennial ryegrass, tall fescue (*Festuca arundinacea* Schreb.), and warm-season grasses, however, generally appear to be resistant to summer patch in the field (8).

Symptoms

Outbreaks of summer patch usually occur from June through September (39). The disease is favored when sustained, high temperatures (25-30 C) are accompanied or followed by periods of heavy rainfall (17, 35). Although symptoms may be exacerbated by hot, dry weather, drought stress is usually not a predisposing factor in this disease (16). Summer patch typically occurs in mature stands of turf three or more years of age, but has occasionally been observed in younger stands (7). Patches may continue to expand the following growing season if conditions are conducive for disease development (35).

On Kentucky bluegrass turf, gray-green, wilting plants initially appear in poorly delineated patches 3-8 cm in diameter (Plates 25-27). White banded lesions (Plate 33) may also be present on leaf blades during periods of heat stress. Circular or irregularly shaped patches, rings, and serpentine patterns, 30 cm or more in diameter (Plates 29-32), later become apparent as the withered leaves turn tan or brown (16, 35).

On golf greens, summer patch symptoms somewhat resemble those of take-all patch on bentgrass turf. Annual bluegrass plants infected with *M. poae* turn yellow in small circular patches that increase in diameter from 5-30 cm (Plate 36). The seemingly spontaneous development of large patches, however, is not uncommon (22). In both situations, the diseased plants progress from yellow to brown, and the turf is killed or severely thinned (Plate 28). As infection centers enlarge and coalesce, large areas of turf may become damaged (22, 39). Bentgrass or weeds often recolonize diseased areas as annual bluegrass declines (Plate 35).

On fairways and lawns, distinct circular patches can occur. Frequently, however, symptoms may appear as little more than diffuse patterns (Plate 34) that are easily confused with other turf disorders such as heat stress, insect damage, or other diseases (35).

Field Diagnosis

Examination of infected roots and crowns may reveal the presence of sparse, ectotrophic hyphae (Plates 37 and 39) that are occasionally aggregated into mycelial mats on tissue surfaces (23). The presence of vascular discoloration in root and crown tissue, however, is of greater diagnostic significance (Plate 38). A good hand lens, and preferably a binocular microscope, is needed for inspection of mycelial characteristics or vascular discoloration. Fruiting structures have not been observed under field conditions (35).

The diagnosis of summer patch in the field frequently poses a problem because, in many situations, typical patch symptoms are absent, vascular discoloration is difficult to discern, and signs of the fungus are sparse or lacking. Prevailing environmental conditions (high temperatures and excessive rainfall), and the host preference of *M. poae* (Kentucky bluegrass, annual bluegrass, and fine-leaved fescues) may be the only indicators available to suggest that the damage is attributable to summer patch. As a result, positive diagnosis can only be confirmed when infected turf is sent to a laboratory for analysis.

SPRING DEAD SPOT

The name "spring dead spot" was coined in 1960 by Wadsworth and Young (42) to describe a disease of bermudagrass (*Cynodon dactylon* (L.) Pers.) in Oklahoma. Observed sporadically since 1936 (42), spring dead spot is now recognized as a major disease of mature bermudagrass turf throughout North America (Fig. 2.3) where winters are sufficiently cold to induce dormancy of the grass host (24).

During the 1960s, a similar disease of bermudagrass (couch grass) was reported in Australia (36). This disease, also called spring dead spot, was later associated with two fungal pathogens from the genus *Leptosphaeria* Ces. & De Not., nom. cons. (37, 45). One of the species, *L. narmari* J. C. Walker & A. M. Sm., the more common incitant of spring dead spot in Australia (45), has also been identified as a causal agent of spring dead spot on bermudagrass in New Zealand (15). The other species, *L. korrae* J. C. Walker & A. M. Sm., is now recognized as an incitant of spring dead spot in Australia and in the United States (6, 13, 45). *L. korrae* has also been established as the causal agent of necrotic ring spot in the United States (34, 46, 47).

Other pathogens associated with spring dead spot of bermudagrass in North America include: *Ophiosphaerella herpotricha* (Fr.:Fr.) J. C. Walker in Kansas, Kentucky, Louisiana, Missouri, North Carolina, Oklahoma, and Texas (30, 40); and *G. graminis* (Sacc.) Arx & D. Olivier var. *graminis* in North Carolina (26). *Rhizoctonia* and *Pythium* species have been linked to spring dead spot symptoms on *Zoysia* spp. in Japan (18).

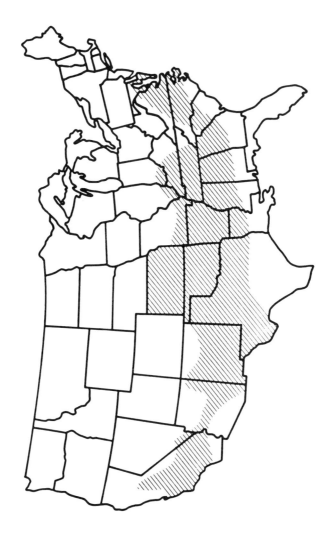

Figure 2.3. Geographical distribution of spring dead spot in the United States.

Host Range

The four fungi primarily associated with spring dead spot (*L. korrae, L. narmari, O. herpotricha,* and *G. graminis* var. *graminis*) are pathogenic in varying degrees on several cool- and warm-season grasses in North America, Europe, and Australia (39). Grasses from which these fungi have been isolated include bermudagrass, magennis bermudagrass (*C. magennisii* (Hurcombe)), African bermudagrass (*C. transvaalensis* Burtt-Davy), tropical carpetgrass (*Axonopus compressus* (Sw.) P. Beauv.), St. Augustinegrass (*Stenotaphrum secundatum* (Walter) Kuntze), kikuyugrass (*Pennisetum clandestinum* Hochst. ex Chiov.), and *Zoysia* spp. (35, 36, 39, 42). Further investigation is required, however, to establish the distribution and host range of these (and perhaps other) ectotrophic root-infecting fungi in regions where typical spring dead spot symptoms occur.

Symptoms

The fungi that cause spring dead spot appear to grow most actively in the fall and spring during cool, moist weather (8, 35). On severely infected plants, the foliage is straw-colored or bleached, and the decaying stolons are blackened (Plate 4). Well defined, roughly circular patches of dead grass appear when the growth of bermudagrass resumes in the spring (Plates 1 and 3). These patches vary in size from 5 cm to more than 1 m in diameter and enlarge as symptoms recur in the same location each year (42). After several years, the disease may subside and symptoms disappear completely (29).

The recovery of turf from spring dead spot is generally a slow process (8, 26). Diseased areas may be recolonized by the growth of stolons from plants surviving in the center or periphery of affected patches. In the southcentral United States, several seasons may elapse before a full turf cover is achieved (42). In the southeast, however, complete recovery may occur within one growing season (8, 39). Where recovery is slow, roots arising from stolon nodes are discolored and malformed, suggesting the presence of a phytotoxin in the soil or in decomposing plant residues that dissipates with time. As a result, areas of slightly depressed and thinned turf may still be visible at the end of the summer (39).

Extensive fungal involvement is readily noticeable on affected bermudagrass wherever spring dead spot occurs. Dark brown mycelia, typically associated with dead and dying roots, crowns, and stolons, comprise a rich flora of dematiaceous fungi that, along with other fungi, apparently function as secondary invaders. Such saprophytic fungi are particularly common on turf affected by spring dead spot in the United States, a situation that no doubt has contributed to the confusion surrounding the etiology of this disease in North America (39).

Signs of *L. korrae* and *L. narmari* are evident on infected plants as dark brown mycelia, mycelial aggregates (plate mycelia), and flattened, lens-shaped sclerotia. Pseudothecia of these fungi, present between the dead leaf sheaths and superficially on decayed stolons, are of common occurrence in Australia (45). Fruiting bodies, however, have only occasionally been seen under field conditions in the United States (N. Jackson and P. H. Dernoeden, *unpublished data*).

O. herpotricha produces dark-colored, ectotrophic mycelia, occasionally with bulbil-like structures, on roots and stolons (40). Pseudothecia, partially immersed in dead leaf sheath and stolon tissues, have recently been observed on bermudagrass affected with spring dead spot in Louisiana (N. Jackson, *unpublished data*).

G. graminis var. *graminis* also produces dark-colored, ectotrophic mycelia. In contrast to the other incitants of this disease, however, *G. graminis* var. *graminis* produces lobed hyphopodia that are formed abundantly on the surfaces of living leaf sheaths and stolons. Perithecia of this fungus are often formed between diseased leaf sheaths and, like hyphopodia, are visible with a hand lens (39).

28

Field Diagnosis

Spring dead spot typically appears in the spring during periods of cool, wet weather as dormant bermudagrass resumes growth. The disease usually occurs on intensively managed turfs that are two or more years old (26). Spring dead spot is apparent as variable-sized, dead patches of turf that heal slowly, recur annually, and are often colonized by unsightly weed grasses and broad-leaved weeds (26) (Plate 2). Straw-colored and bleached foliage, accompanied by blackened, rotted roots and stem bases, is a common diagnostic feature in the field. Sundry species of dark-colored fungi may be present in and around necrotic tissues. Fruiting structures of the fungi currently known to cause spring dead spot are often absent. In such cases, accurate identification cannot be made without appropriate laboratory analysis.

NECROTIC RING SPOT

During the past decade, another patch disease resembling take-all patch has occurred with increasing frequency primarily on Kentucky bluegrass turf in the upper midwest, the Pacific northwest, and in the northeastern regions of the United States (Fig. 2.4) (8). In 1983, Worf et al (46) named this disease, then of undetermined cause, necrotic ring spot. Concurrent research by turf pathologists at several locations in the early 1980s established *L. korrae* as the incitant of this disease (2, 33, 34, 47). Severe damage has been most frequently reported on Kentucky bluegrass lawns and sports turf in the early years following establishment from sod. The disease can affect other cool-season turfgrasses and may also appear in younger, seeded swards (3).

Host Range

L. korrae is a proven pathogen on a number of cool- and warm-season turfgrass and cereal hosts (45). Confirmed reports

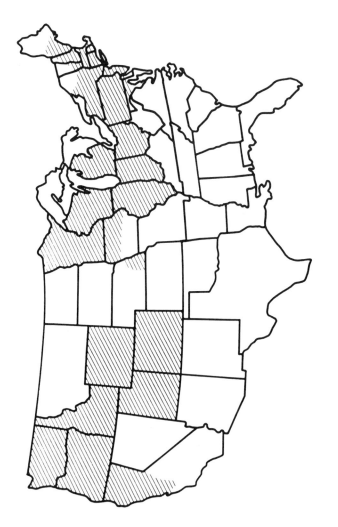

Figure 2.4.   Geographical distribution of necrotic ring spot in the United States.

of necrotic ring spot in the field are presently confined to Kentucky bluegrass, rough bluegrass, annual bluegrass, and red fescue (*F. rubra* L.). The causal agent has also been isolated from carpetgrass (*Axonopus* spp.), bermudagrass, and centipedegrass (*Eremochloa ophiuroides* (Munro) Hack.) (35).

Symptoms

Symptoms of necrotic ring spot may appear in the spring or late summer during periods of cool, wet weather. Patches often fade with the advent of warmer temperatures in the summer, but may reappear in response to heat or drought stress. Symptoms that develop in the fall may remain visible through the winter (35). Even where patch symptoms were mild the previous growing season, plants on these sites are late to break dormancy and appear stunted when compared to surrounding, healthy turf (39).

Patches of turf infected with *L. korrae* range in size from 5-30 cm or more in diameter (Plates 7, 8, and 10). The older, outer leaves of infected grass plants turn successively yellow, brown, and may appear bleached. Inner leaves initially develop a bronze or purple to wine-red cast (Plate 9) and later turn brown as tillers and whole plants succumb to infection (3, 34, 39, 47). In turf containing a thick thatch layer, plants within these patches may collapse to form crater-like depressions 2-3 cm in depth (Plate 5). In other situations, plants located in the center of patches may survive, resulting in a ring or frogeye appearance (Plate 6). As the disease progresses and the infection centers coalesce, large areas of turf may become thinned or severely scarred (39, 47).

Infected plants have little root hold as roots, rhizomes, and crowns are progressively invaded by the fungus. These tissues blacken as they become infected with dark-colored, ectotrophic, hyphae that often form aggregates of plate mycelia (Plates 11 and 12). Black pseudothecia may be found singly or in groups on necrotic roots, rhizomes, leaf sheaths, and crowns (Plates 13-15). Fruiting is sporadic, however, and is restricted to tissues that have been dead for an extended period of time (i.e., 2-3 mo) (2, 34, 39).

Severe deterioration of the turf is associated with the annual encroachment of the fungus from the perimeter of existing patches and with the appearance of new infection centers. After a few years, however, symptoms may subside. This phenomenon is similar to that of take-all decline (32).

Field Diagnosis

In Kentucky bluegrass, the development of sunken, crater-like patches during cool, moist weather and the presence of ectotrophic, dark-colored mycelia associated with blackened roots, rhizomes, and stem bases are good indicators of necrotic ring spot. The disease is most prevalent two to four years after sodding, although symptoms may also appear in newly seeded areas. Pseudothecia, visible with a hand lens but often hard to find, are a conclusive diagnostic feature.

The pattern of necrotic ring spot development on Kentucky bluegrass turf is similar to that of Fusarium blight (5) and summer patch (22), but symptoms are initiated at cooler temperatures (47). Conditions conducive to the development of necrotic ring spot are more akin to those favoring yellow patch, caused by *Rhizoctonia cerealis* Van der Hoeven (1). In fact, necrotic ring spot closely resembles yellow patch in the field and the two diseases may have been confused on numerous occasions in the past.

BERMUDAGRASS DECLINE

Bermudagrass decline is a newly described disease of bermudagrass maintained as golf course putting greens (10). *G. graminis* var. *graminis*, the primary fungus associated with this disease, has also been implicated as a causal agent of spring dead spot in North Carolina (25, 26). Other ectotrophic fungi, such as *G. incrustans* Landschoot & Jackson and *Phialophora* spp., have been associated with declining bermudagrass turf.

These fungi, however, appear to be either saprophytes or secondary pathogens that may contribute to an overall decline complex (10, 12, 35).

Host Range

Bermudagrass decline has been observed on bermudagrass spp. and bermudagrass hybrids (*C. dactylon* X *C. transvaalensis*) in Florida and in other regions in the southeastern United States (10, 12). *G. graminis* var. *graminis* has been isolated from symptomatic roots of bermudagrass in Australia (43) and from African bermudagrass in Australia and South Africa (31, 43). The pathogenicity of *G. graminis* var. *graminis*, obtained from St. Augustinegrass in Florida and Alabama, has been confirmed on this in host greenhouse studies (11). To date, bermudagrass decline has only been associated with closely-mowed, bermudagrass putting greens, although the pathogen has been isolated from asymptomatic turf in higher-mowed areas. All known cultivars of bermudagrass are susceptible to this disease (35). Tifdwarf and Tifgreen, hybrid bermudagrass cultivars widely used on golf course greens in Florida, are particularly susceptible (12).

Symptoms

In the southeastern United States, bermudagrass decline commonly occurs during the summer and fall months when the majority of annual precipiataion is received and the weather is hot and humid (10). The disease first appears in the field as irregularly shaped patches of chlorotic turf that vary from 0.2-1 m in diameter. Generally, the outer margins of golf greens are the first to exhibit symptoms of this disease (12) (Plate 66).

On individual plants, the lower leaves are the first to become chlorotic and die (Plate 67). As the chlorosis spreads into the upper foliage, dark-colored lesions often form on the roots. Roots, stolons, and rhizomes eventually blacken and decay (Plate 68). In the later stages of infection, patches often coalesce into large areas of blighted turf (12). Dead patches may be recolonized by new bermudagrass tillers, but these plants may also become infected and die (35).

Field Diagnosis

The symptoms of bermudagrass decline may be confused with other diseases caused by *Pythium* spp. and nematodes (35). Diagnostic features of bermudagrass decline include a foliar chlorosis that begins on the lower leaves and progresses into the upper foliage. Roots, stolons, and rhizomes become dark brown to black and decay (12).

G. graminis var. *graminis* forms dark brown to black ectotrophic runner hyphae on roots, stolons, and rhizomes of bermudagrass turf. The fungus penetrates host tissue through infection hyphae that arise from lobed hyphopodia. Prolonged incubation of infected plant material is usually required to encourage the development of perithecia (35). Where apparent, perithecia are dark brown to black, flask-shaped structures with curved necks that often protrude through the leaf sheaths of infected plants (12, 44). Although phialospores of *G. graminis* var. *graminis* have been produced on an agar medium (35), the anamorph has not been observed on infected tissue in the field.

ACKNOWLEDGMENTS

I wish to acknowledge and thank Dr. Peter Dernoeden and Dr. Peter Landschoot for providing unpublished data and other information presented in this review. Dr. Landschoot also graciously compiled the figures in this chapter in association with Drs. Gary A. Chastagner, Bruce B. Clarke, Peter H. Dernoeden, John R. Hartman, Noel Jackson, Bobby Joyner, Leon T. Lucas, Eric B. Nelson, Gail L. Schumann, Donald Scott, William W. Shane, Ned A. Tisserat, Joseph M. Vargas, John Watkins, Henry T. Wilkinson, and Gayle L. Worf.

LITERATURE CITED

1. Burpee, L. 1980. *Rhizoctonia cerealis* causes yellow patch of turfgrasses. Plant Dis. 64:1114-1116.
2. Chastagner, G. A., Goss, R. L., Staley, J. M., and Hammer, W. 1984. A new disease of bluegrass turf and its control in the Pacific Northwest. (Abstr.) Phytopathology 74:811-812.
3. Chastagner, G. A., and Hammer, B. 1987. Current research on necrotic ring spot. Pages 94-95 in: Proc. Northwest Turfgrass Conf., 41st. Northwest Turfgrass Association, Gleneden Beach, OR.
4. Christensen, M. J. 1985. Turf Diseases Management-- New Zealand. Pages 106-111 in: Proc. N. Z. Sports Turf Convention, 3rd. Murphy, J. W., ed. Grasslands Division, Palmerston North.
5. Couch, H. B., and Bedford, E. R. 1966. Fusarium blight of turfgrasses. Phytopathology 56:781-786.
6. Crahay, J. N., Dernoeden, P. H., and O'Neill, N. R. 1988. Growth and pathogenicity of *Leptosphaeria korrae* in bermudagrass. Plant Dis. 72:945-949.
7. Davis, D. B., and Dernoeden, P. H. 1991. Summer patch and Kentucky bluegrass quality as influenced by cultural practices. Agron. J. 83:670-677.
8. Dernoeden, P. H. 1989. Symptomatology and management of common turfgrass diseases in transition zone and northern regions. Pages 273-296 in: Integrated Pest Management for Turfgrass and Ornamentals. A. R. Leslie and R. L. Metcalf, eds. U. S. Environmental Protection Agency, Washington, D. C.
9. Dernoeden, P. H., and O'Neill, N. R. 1983. Occurrence of Gaeumannomyces patch disease in Maryland and growth and pathogenicity of the casual agent. Plant Dis. 67:528-532.
10. Elliott, M. L. 1991. Determination of an etiological agent of bermudagrass decline. Phytopathology 81:1380-1384.

11. Elliott, M. L., Hagan, A. K., and Mullen, J. M. 1993. Association of *Gaeumannomyces graminis* var. *graminis* with a St. Augustinegrass root rot disease. Plant Dis. 77:206-209.

12. Elliott, M. L., and Landschoot, P. J. 1991. Fungi similar to *Gaeumannomyces* associated with root rot of turfgrasses in Florida. Plant Dis. 75:238-241.

13. Endo, R. M., Ohr, H. D., and Krausman, E. M. 1985. *Leptosphaeria korrae*, a cause of the spring dead spot disease of bermudagrass in California. Plant Dis. 69:235-237.

14. Jackson, N. 1980. Gaeumannomyces (Ophiobolus) patch disease. Pages 141-143 in: Advances in Turfgrass Pathology. P. O. Larsen and B. G. Joyner, eds. Harcourt Brace Jovanovich, Duluth, Mn.

15. Jackson, N. 1987. Some patch diseases of turfgrasses associated with root and crown infecting fungi. N. Z. J. Turf Management 1:21-23.

16. Kackley, K. E., Grybauskas, A. P., Dernoeden, P. H., and Hill, R. L. 1990. Role of drought stress in the development of summer patch in field-inoculated Kentucky bluegrass. Phytopathology 80:655-658.

17. Kackley, K. E., Grybauskas, A. P., Hill, R. L., and Dernoeden, P. H. 1990. Influence of temperature-soil water status interactions on the development of summer patch in *Poa* spp. Phytopathology 80:650-655.

18. Kawanabe, Y. 1991. The occurrence and control of turfgrass diseases in Japan. Jpn. Pestic. Inf. 59:5-9.

19. Kemp, M. L. 1991. The susceptibility of fine fescues to isolates of *Magnaporthe poae* and *Gaeumannomyces incrustans*. Ph.D. dissertation. Rutgers University, New Brunswick, NJ.

20. Kemp, M. L., Clarke, B. B., and Funk, C. R. 1990. The susceptibility of fine fescues to isolates of *Magnaporthe poae* and *Gaeumannomyces incrustans*. (Abstr.) Phytopathology 80:978.

21. Kemp, M. L., Landschoot, P. J., Clarke, B. B., and Funk, C. R. 1990. Response of fine fescues to field inoculation with summer patch. Page 176 in: Agronomy Abstracts. American Society of Agronomy, Madison, WI.

22. Landschoot, P. J., Clarke, B. B., and Jackson, N. 1989. Summer patch in the northeast. Golf Course Management 57:38-42.

23. Landschoot, P. J., and Jackson, N. 1989. *Magnaporthe poae* sp. nov., a hyphopodiate fungus with a *Phialophora* anamorph from grass roots in the United States. Mycol. Res. 93:59-62.

24. Lucas, L. T. 1980. Spring deadspot of bermudagrass. Pages 183-187 in: Advances in Turfgrass Pathology. P. O. Larsen and B. G. Joyner, eds. Harcourt Brace Jovanovich, Duluth, MN.

25. McCarty, L. B., and Lucas, L. T. 1988. Identification and suppression of spring dead spot disease in bermudagrass. Page 154 in: Agronomy Abstracts. American Society of Agronomy, Madison, WI.

26. McCarty, L. B., and Lucas, L. T. 1989. *Gaeumannomyces graminis* associated with spring dead spot of bermudagrass in the southeastern United States. Plant Dis. 73:659-661.

27. Nilsson, H. E. 1969. Studies of root and foot rot diseases of cereals and grasses. I. On resistance to *Ophiobolus graminis* Sacc. Lantbrukshögsk. Ann. 35:275-807.

28. Nilsson, H. E., and Smith, J. D. 1981. Take-all of grasses. Pages 433-448 in: Biology and Control of Take-all. M. J. C. Asher and P. J. Shipton, eds. Academic Press, New York.

29. Pair, J. C., Crowe, F. J., and Willis, W. G. 1986. Transmission of spring dead spot disease of bermudagrass by turf/soil cores. Plant Dis. 70:877-878.

30. Sauer, K. M., Hulbert, S. H., and Tisserat, N. A. 1993. Identification of *Ophiosphaerella herpotricha* by cloned DNA probes. Phytopathology 83:97-102.

31. Scott, D. B. 1989. *Gaeumannomyces graminis* var. *graminis* on Gramineae in South Africa. Phytophylactica 21:251-254.

32. Shipton, P. J. 1975. Take-all decline during cereal monoculture. Pages 137-144 in: Biology and Control of Soilborne Plant Pathogens. G. W. Bruehl, ed. American Phytopathological Society, St. Paul, MN.

33. Smiley, R. W. 1984. "Fusarium blight syndrome" re-described as a group of patch diseases caused by *Phialophora graminicola*, *Leptosphaeria korrae*, or related species. (Abstr.) Phytopathology 74:811.

34. Smiley, R. W., and Craven Fowler, M. C. 1984. *Leptosphaeria korrae* and *Phialophora graminicola* associated with Fusarium blight syndrome of *Poa pratensis* in New York. Plant Dis. 8:440-442.

35. Smiley, R. W., Dernoeden, P. H., and Clarke, B. B. 1992. Compendium of Turfgrass Diseases. 2nd ed. American Phytopathological Society, St. Paul, MN.

36. Smith, A. M. 1965. *Ophiobolus herpotrichus*, a cause of spring dead spot in couch turf. Agric. Gaz. N. S. W. 76:753-758.

37. Smith, A. M. 1971. Spring dead spot of couch grass turf in New South Wales. J. Sports Turf Res. Inst. 47:54-59.

38. Smith, J. D. 1956. Fungi and turf diseases. (6) Ophiobolus patch disease. J. Sports Turf Res. Inst. 9:180-202.

39. Smith, J. D., Jackson, N., and Woolhouse, A. R. 1989. Fungal Diseases of Amenity Turf Grasses. E. & F. N. Spon, London.

40. Tisserat, N. A., Pair, J. C., and Nus, A. 1989. *Ophiosphaerella herpotricha*, a cause of spring dead spot of bermudagrass in Kansas. Plant Dis. 73:933-937.

41. von Arx, J. A., and Olivier, D. L. 1952. The taxonomy of *Ophiobolus graminis* Sacc. Trans. Br. Mycol. Soc. 35:29-33.

42. Wadsworth, D. F., and Young, H. C., Jr. 1960. Spring dead spot of bermudagrass. Plant Dis. Rep. 44:516-518.

43. Walker, J. 1972. Type studies on *Gaeumannomyces graminis* and related fungi. Trans. Br. Mycol. Soc. 58:427-457.

44. Walker, J. 1980. *Gaeumannomyces, Linocarpon, Ophiobolus* and several other genera of scolecospored ascomycetes and *Phialophora* conidial states, with a note on hyphopodia. Mycotaxon 11:1-129.

45. Walker, J., and Smith, A. M. 1972. *Leptosphaeria narmari* and *L. korrae* spp. nov., two long-spored pathogens of grasses in Australia. Trans. Br. Mycol. Soc. 58:459-466.

46. Worf, G. L., Brown, K. J., and Kachadoorian, R. V. 1983. Survey of "necrotic ring spot" disease in Wisconsin lawns. (Abstr.) Phytopathology 73:839.

47. Worf, G. L., Stewart, J. S., and Avenius, R. C. 1986. Necrotic ring spot disease of turfgrass in Wisconsin. Plant Dis. 70:453-458.

TAXONOMY AND BIOLOGY OF ECTOTROPHIC ROOT-INFECTING FUNGI ASSOCIATED WITH PATCH DISEASES OF TURFGRASSES

Peter J. Landschoot
The Pennsylvania State University
University Park, PA 16802

INTRODUCTION

Patch diseases, caused by ectotrophic root-infecting (ERI) fungi, represent a newly recognized and highly destructive group of turfgrass diseases in the United States. Although some of these fungi have been studied extensively in Europe and Australia, most were not identified as pathogens of turf in the United States prior to 1984 (41). Knowledge of the taxonomy and biology of the ERI fungi is important to plant disease diagnosticians and to those involved in devising strategies for disease control. These fungi require special techniques for their isolation and identification, and are, therefore, unfamiliar to many plant pathologists (43). This chapter briefly reviews the biology and taxonomy of the ERI fungi associated with patch diseases. Special emphasis has been placed on species descriptions and the morphological structures and diagnostic procedures used to identify these fungi.

CLASSIFICATION OF THE ECTOTROPHIC ROOT-INFECTING FUNGI

Patch diseases of turfgrasses are caused by a closely related group of ERI fungi. All are classified in either the Ascomycotina or the Deuteromycotina. The classification scheme in Table 3.1 has been adapted from Ainsworth et al (1), Alexopoulos and Mims (2), Farr et al (17), and Schol-Schwarz (36).

DIAGNOSTIC FEATURES

The ERI fungi possess several morphological features that are useful taxonomic aids. Because many of these fungi are quite similar, multiple characteristics are often required to distinguish one ERI fungus from another (51).

Sexual Structures

Structures associated with the teleomorph represent the most useful taxonomic aids for the identification of the Ascomycetes (1). Diagnostic features of the ascocarp include its position on the host or substrate; length and width of the neck; presence of paraphyses; color, size, shape, and thickness of the wall; and the surface texture. Characteristics of the ascus that may aid in identification include its size, shape, and position in the ascocarp. Ascospore morphology is also important when classifying members of the ERI fungi. Features such as ascospore size, shape, color (*en masse*), number of septations, arrangement in the ascus, and mode of germination all need to be considered when attempting to identify unknown isolates (53).

Perithecia

Gaeumannomyces and *Magnaporthe* are members of the Hymenoascomycetidae I (Table 3.1). They are characterized by the presence of ascocarps (perithecia) that are entirely enclosed

Table 3.1.   Classification of Ectotrophic root-infecting fungi associated with patch diseases of turfgrasses.

Division: Amastigomycota

Subdivision: Ascomycotina
  Class: Ascomycetes

   Subclass: Hymenoascomycetidae I (Pyrenomycetes)
     Order: Diaporthales (Sphaeriales)
       Family: Diaporthaceae
         Genus: *Gaeumannomyces*
           Species: *Gaeumannomyces graminis* var. *avenae*
                          Gaeumannomyces graminis var. *graminis*
                          Gaeumannomyces incrustans
                          Gaeumannomyces cylindrosporus
         Genus: *Magnaporthe*
           Species: *Magnaporthe poae*

   Subclass: Loculoascomycetidae (Loculoascomycetes)
     Order: Dothidiales (Pleosporales)
       Family: Pleosporaceae
         Genus: *Leptosphaeria*
           Species: *Leptosphaeria korrae*
                          Leptosphaeria narmari
         Genus: *Ophiosphaerella*
           Species: *Ophiosphaerella herpotricha*

Subdivision: Deuteromycotina
  Form-Class: Deuteromycetes
    Form-Subclass: Hyphomycetidae (Hyphomycetes)
      Form-Order: Hyphomycetales (Moniliales)
        Form-Family: Moniliaceae
          Form-Genus: *Phialophora*
            Form-Species: *Phialophora graminicola*
                              Phialophora radicicola

---

within a peridial wall. The perithecia are predominantly flask-shaped (Plates 19, 43, 61, 62, and 63) and contain an ostiole lined with slender, hyaline periphyses. Asci are typically clavate in shape, unitunicate (possessing only one ascus wall), and are arranged in a hymenial layer (Plates 20, 44, and 63). In both genera, the ascus apex is surrounded by a refractive ring. This ring absorbs cotton blue stain but does not turn blue when stained with iodine (33).

Pseudothecia

The ERI fungi classified in the Loculoascomycetidae (Table 3.1) (i.e., *Leptosphaeria* and *Ophiosphaerella*) possess ascocarps that form within stromal tissue or are grouped in locules. The mature ascocarp or ascostroma is often referred to as a pseudothecium and is partially or entirely immersed in host tissue (Plates 13-15). The genera in this subclass are characterized by the presence of bitunicate asci, which possess a thin outer wall and a thick inner wall. These asci, however, lack refractive apical rings (30).

Asexual Structures

Conidial State

Several of the ERI fungi have a *Phialophora* anamorph (Table 3.1). Although differences in the morphology of the conidial state exist among species of the ERI fungi, difficulties may arise when attempting to distinguish between species using this feature alone. Schol-Schwarz (36) cautioned that considerable variation exists within *Phialophora* conidial states. The author also noted that, even within the same species, variation in the morphology of conidia and conidiophores is often dependent on the age of the culture, the growth medium, and the intensity and duration of the light source. When growth parameters are specified, however, the use of multiple asexual characters (i.e., size, shape, color, and arrangement of conidia; size, shape, and color of conidiophores; and the size, shape, and arrangement of the conidiogenous cells) may be useful in identifying species of the ERI fungi (53).

Hyphopodia

Walker (52, 53) defined a hyphopodium as an organ of attachment and penetration that is produced from vegetative, epiphytic hyphae on host tissue. Hyphopodia may occur in a terminal, lateral, or intercalary position. They may be simple or lobed, hyaline or pigmented, and sessile or stalked (Plates 40, 41, 60, 69, and 70). Differences in hyphopodial morphology are often used to distinguish between varieties of *Gaeumannomyces graminis* (Sacc.) Arx & D. Olivier and some *Phialophora* spp. (51, 52). Hyphopodia are usually found on the surfaces of stems and leaf sheaths. Occasionally, hyphopodia develop on root surfaces or within host tissue (11, 52). In culture, hyphopodia of *Gaeumannomyces* spp., *Magnaporthe poae* Landschoot & Jackson, and *Phialophora graminicola* (Deacon) J. Walker have even been observed adhering to the sides of petri dishes (P. J. Landschoot, *personal observation*).

Currently, there is some disagreement within the scientific community regarding the use of the term hyphopodium. Traditionally, adhesion and penetration structures produced on germtubes have been referred to as appressoria, whereas similar structures produced on vegetative hyphae have been defined as hyphopodia (2, 15). In a review of the concept, Emmett and Parbery (15) classified all adhesion and penetration structures on the basis of function rather than morphology. Using this approach, any structure used by a fungus to gain entry into a host would be defined as an appressorium. Walker (53) noted the similarity between appressoria and hyphopodia but suggested retaining the term hyphopodium because of its usefulness in distinguishing between different types of hyphal outgrowths. In this chapter, the term hyphopodium will be used because of its taxonomic significance.

Growth Cessation Structures

Darkly-pigmented swollen cells, similar in appearance to hyphopodia, have been observed in roots infected with ERI fungi (7, 8, 9, 37, 53) (Plate 42). Deacon (8) called these cells "growth cessation structures." The author speculated that these structures form when hyphal penetration is slowed or stopped by the host. Although growth cessation structures are not frequently used as taxonomic characters, Deacon and Scott (9) noted their potential as a distinguishing feature for some *Phialophora* spp.

Cultural Characteristics

Cultural characteristics can be useful in the identification of the ERI fungi. In culture, species of ERI fungi may vary in colony color, growth habit (i.e., aerial or appressed mycelium; single or aggregated hyphal growth), and in the production of stromatic structures (53). Since these features may change depending on the age of the culture or the conditions under which it is stored, colony morphology should be observed frequently and under different environmental conditions. For example, the growth rate of many ERI fungi varies dramatically at different temperatures. During a period of 24 h, *M. poae* may attain a colony radius of 7-8 mm on potato dextrose agar at a temperature of 30 C, whereas *P. graminicola* grows little or not at all at this temperature (26). As a result, comparable growth parameters must be used when attempting to differentiate between genera, species, or varieties of the ERI fungi.

TAXONOMIC DESCRIPTIONS OF THE CAUSAL AGENTS

The ERI fungi associated with turfgrass patch diseases include members of the genera *Gaeumannomyces*, *Magnaporthe*, *Leptosphaeria*, *Ophiosphaerella*, and the form-genus *Phialophora*. Detailed descriptions of the individual species within these genera can be found in Cain (3); Landschoot and Jackson (27, 28); Walker (51, 52, 53); Walker and Smith (54); and Webster and Hudson (55). Smiley et al (41) and Walker (53) have also developed diagnostic keys for the provisional identification of the ERI fungi. A concise summary of the taxonomy, sexuality, and methodology used to identify these fungi is presented in the remainder of this chapter and in Tables 3.2a, 3.2b, 3.3a, and 3.3b.

Gaeumannomyces Arx & D. Olivier

Gaeumannomyces species possess ostiolate perithecia with long necks and upwardly pointing periphyses. All species have unitunicate asci with a refractive ring at the apex. Ascospores are typically hyaline and have numerous septa (53). Another important feature of this genus is the presence of superficial hyphae in the form of hyphal strands, radiating fans, or hyphopodia (52). Species within the genus *Gaeumannomyces* are parasitic on members of the Gramineae and Cyperaceae (46, 53).

Gaeumannomyces graminis (Sacc.) Arx & D. Olivier var. *avenae* (E. M. Turner) Dennis
syn: *Ophiobolus graminis* (Sacc.) Sacc. var. *avenae* E. M. Turner

G. graminis var. *avenae* causes take-all of oats and take-all patch (formerly known as Ophiobolus patch) of turfgrasses. Although take-all patch is usually confined to bentgrass (*Agrostis* L.) species, it may also affect fine-leaved fescue (*Festuca* L.) and bluegrass (*Poa* L.) species (35).

Taxonomy. In the field, perithecia of *G. graminis* var. *avenae* are occasionally found on stem bases and leaf sheaths of infected grasses (41, 43). The perithecia are dark brown to black, flask-shaped (Table 3.2a), and have slightly curved necks (Plate 62). Asci are unitunicate, clavate, contain eight spores,

Table 3.2a. Sexual characteristics of *Gaeumannomyces* species associated with patch diseases of turfgrasses.[z]

Sexual character	G. graminis var. avenae	G. graminis var. graminis	G. incrustans	G. cylindrosporus
Ascocarp				
Body	300-500 X 250-400	200-400 X 150-300	179-420	(250)280-565(600)
Neck:				
length	100-400	100-400	252-672	< 250
width	70-100	70-100	84-147	48-116(120)
Asci				
length	(90)110-150(160)	(80)100-135(140)	65-107	(60)65-135
width	12-16	10-15	6-11	9-16
wall	unitunicate	unitunicate	unitunicate	unitunicate
Ascospores				
length	(85)100-130(140)	(70)80-105(110)	33-53	(35)37-69(75)
width	(2)2.5-3.5(4)	2-3(4)	2-4	3-5(6)
septa	(3)5-11(15)	3-5(8)	3-5	3-4(8)
Color (*en masse*)	yellowish	yellowish	yellowish	yellowish

[z]All measurements are in μm. Dimensions that fall beyond the normal range for each character have been placed in parentheses. Information in this table has been compiled from references 27, 43, 52, and 53.

Table 3.2b. Sexual characteristics of several ectotrophic root-infecting fungi associated with patch diseases of turfgrasses.[z]

Sexual character	*Magnaporthe poae*	*Leptosphaeria korrae*	*L. narmari*	*Ophiosphaerella herpotricha*
Ascocarp				
Body	252-556	300-550 X 300-500	500-700 X 650	(250)300-400
Neck:				
length	357-756	50-150(200)	100-300	100(200)
width	95-179	200(250)	300-450	80
Asci				
length	63-108	(145)150-185(200)	(100)110-145(155)	(110)150-190(218)
width	7-15	10-13(15)	11-13	7-9(11)
wall	unitunicate	bitunicate	bitunicate	bitunicate
Ascospores				
length	23-42	(120)140-170(180)	(35)45-62(72)	(120)140-180(215)
width	4-6	4-5(5.5)	4-5(6)	2-2.5(3)
septa	3	(1-6)7(15)	(3)5(7)	(8)12-16(20)
Color (*en masse*)	pale brown	pale brown	pale brown	pale brown

[z]All measurements are in μm. Dimensions that fall beyond the normal range for each character have been placed in parentheses. Information in this table has been compiled from references 28, 43, 49, 52, 53, 54, and 55.

49

Table 3.3a. Asexual characteristics of *Gaeumannomyces* species associated with patch diseases of turfgrasses[z].

Asexual Character	G. graminis var. avenae	G. graminis var. graminis	G. incrustans	G. cylindrosporus
Conidia (µm)	5-14 X 2-4	5-14 X 2-4	3-6 X 2-3	2.5-7 X 1-3
Hyphopodia	simple (unlobed)	simple or strongly lobed	slightly lobed (rare)	simple and slightly lobed
Maximum growth rate (mm^{-24} h)	6-10 (at 20-25 C)	8-12 (at 20-25 C)	7-14 (at 28-30 C)	4-6 (at 25 C)
Colony morphology	short, aerial mycelium, white, gray, or black; distinct stranding and curling back towards colony center	short, aerial mycelium, white, gray, brown, or black; distinct stranding and curling back towards colony center	mycelium appressed, hyaline, turning gray or black at maturity; crusts form on surface, advancing hyphae curling back towards colony center	felt of aerial mycelium that appears white, tan, or gray on potato dextrose agar

[z]Information in this table has been compiled from references 27, 43, 52, and 53.

Table 3.3b. Asexual characteristics of several ectotrophic root-infecting fungi associated with patch diseases of turfgrasses[z].

Asexual Character	Magnaporthe poae	Leptosphaeria korrae	L. narmari	Ophiosphaerella herpotricha
Conidia (μm)	3-8 X 1-3	none	none	69-116 X 3-4
Hyphopodia	slightly lobed	plate-like sclerotia	plate-like sclerotia	intercalary
Maximum growth rate (mm^{-24} h)	7-12 (at 28-30 C)	4-5 (at 25 C)	4-5 (at 25 C)	3.5-4 (at 20-25 C)
Colony morphology	mycelium appressed, hyaline, turning gray or olive-brown at maturity; thick strands of advancing hyphae curling back towards colony center	felt of aerial mycelium, white, light to dark gray	felt of aerial mycelium, white to buff, darkening as colony ages	mycelium cottony white, turning tan to brown in 3-7 days

[z]Information in this table has been compiled from references 28, 43, 49, 52, 53, 54, and 55.

51

and have a refractive ring at the apex when mature (Plate 63). Ascospores are hyaline to light yellow (*en masse*), slightly curved, and contain three to 15 septa (Plates 64 and 65). Paraphyses are hyaline, septate, and are longer than the asci (43, 50, 52, 53).

The conidia (phialospores) of *G. graminis* var. *avenae* are produced in clusters at the ends of phialides (Table 3.3a). Conidia are hyaline, straight to slightly curved (lunate), and taper to an acute base. Smaller, strongly curved, non germinating conidia may also develop at the apex of phialides (Plate 65). Runner hyphae (4-7 µm wide) are brown, septate, and often colonize the surfaces of roots, rhizomes, and stems of infected plants. Infection hyphae are slender, hyaline, and originate from runner hyphae, simple hyphopodia (7-15 µm long X 4-8 µm wide) (Plate 60), or mycelial mats (43, 50, 53).

In culture, the mycelium of *G. graminis* var. *avenae* is initially white but turns gray or black when mature. The agar surface is often covered with a short, gray, aerial mycelium. Thick hyphal strands frequently develop and curl back towards the center of the colony. Optimal radial growth (6-10 mm$^{-24\ h}$) occurs at 20-25 C (43, 52, 53).

Teleomorph production. Several studies have been conducted to assess the factors that affect vegetative growth and perithecial production of *G. graminis* (4, 19, 20, 23, 34, 47, 56, 57, 58). In general, perithecia have been used in genetic studies, for the production of inoculum, and as a source of virulent isolates (39). For *G. graminis*, perithecial formation is usually inhibited by darkness, becomes infrequent or ceases when cultures are maintained for extended periods, and is less prolific *in vitro* than on dead or living plant tissue (23).

In an extensive seedling infection study using 39 isolates of *G. graminis*, Holden and Hornby (23) compared *in vitro* and *in vivo* methods of inducing the formation of perithecia on cereals. They found that production of perithecia by *G. graminis* var. *avenae* was greater *in vivo* than *in vitro* (89 vs. 33%, respectively). For *G. graminis* (Sacc.) Arx & D. Olivier var. *tritici* J. Walker, however, the percentage of perithecia produced using two *in vitro* techniques (e.g., malt extract broth agar or filter paper) was similar to the *in vivo* method. Of the *in vitro* techniques, *G. graminis* var. *avenae* produced the greatest number of perithecia on "Cracklin' Oat Bran" agar and wheat-seedling agar placed at room temperature in indirect light. Dernoeden and O'Neill (10) were able to induce the formation of perithecia of *G. graminis* var. *avenae* on bentgrass seedlings. Mature perithecia developed after 49 days on plants that were grown in sand culture at 25 C and under low light (155 µE m^{-2} s^{-1}).

Gaeumannomyces graminis (Sacc.) Arx & D. Olivier var. *graminis*

> syns: *Rhaphidophora graminis* Sacc.; *Ophiobolus graminis* (Sacc.) Sacc. in Roum. & Sacc.; *Ophiochaeta graminis* (Sacc.) K. Hara; *Ophiobolus oryzinus* Sacc.; *Linocarpon oryzinum* (Sacc.) Petrak; and *Gaeumannomyces oryzinus* (Sacc.) Schrantz

Until recently, *G. graminis* var. *graminis* was considered a saprophyte of turfgrasses (53, 60). In 1988, however, McCarty and Lucas (31) reported an association between this fungus and spring dead spot of bermudagrass (*Cynodon dactylon* (L.) Pers.) in North Carolina. *G. graminis* var. *graminis* has also been implicated as the causal agent of bermudagrass decline in the southern United States (12, 14). Moreover, its pathogenicity has been confirmed on St. Augustinegrass (*Stenotaphrum secundatum* (Walter) Kuntze) in the greenhouse (13).

Taxonomy. The morphological features of *G. graminis* var. *graminis* are quite similar to those of *G. graminis* var. *avenae*. Ascospores of *G. graminis* var. *graminis*, however, are generally smaller than those of *G. graminis* var. *avenae* (Table 3.2a). Another distinguishing characteristic of *G. graminis* var. *graminis* is the presence of both simple (7-15 µm long X

4-10 μm wide) and strongly lobed hyphopodia (15-40 μm long X 10-25 μm wide) (Plates 69 and 70) that develop on the surfaces of stems and leaf sheaths (43, 52, 53) (Table 3.3a). In culture, the mycelium of *G. graminis* var. *graminis* is at first white but at maturity turns gray, brown, or black. Distinct strands of aerial mycelia form and often curl back toward the center of the colony. *G. graminis* var. *graminis* attains optimum radial growth (8-12 mm$^{-24\,h}$) at 20-25 C (52, 53).

Teleomorph production. Holden and Hornby (23) determined that approximately 50% of the *G. graminis* var. *graminis* isolates tested in a wheat (*Triticum aestivum* L.) seedling assay yielded mature perithecia. The *in vitro* methods tested, however, were less effective. Of these, the filter paper method yielded the highest percentage of mature perithecia (36%).

Speakman (48) described a successful method for the production of mature perithecia of *G. graminis* var. *graminis* on malt extract peptone agar fortified with penicillin G (40 μg/ml), streptomycin sulfate (50 μg/ml), and terramycin (oxytetracycline; 60 μg/ml). Following inoculation with the fungus, plates were incubated in the dark at 20 C until the agar surface was completely colonized. Sterile distilled water was then added to each culture and the plates were placed in a north-facing window at 20-22 C. Mature perithecia were observed after 30 days.

Perithecia of *G. graminis* var. *graminis* have also been produced using a modification of another method developed by Speakman (47). In the revised protocol, surface-disinfested wheat seedlings were placed on plates of water agar and inoculated with the fungus. Cultures were then sealed with parafilm and placed in a west-facing window for 6-8 wk or until mature perithecia were produced. This method has also been used to produce conidia and hyphopodia of *G. graminis* var. *graminis* (P. L. Landschoot, *unpublished data*).

Gaeumannomyces incrustans Landschoot & Jackson

G. incrustans is the only heterothallic member of the *Gaeumannomyces* genus (27). Since it has only recently been described as a new species, very little is known about the ecology, distribution, and host range of this fungus. Landschoot (26) found that *G. incrustans* was moderately pathogenic on bluegrass species maintained in the greenhouse at high (28 C) temperatures. In 1990, Kemp et al (25) confirmed the pathogenicity of this fungus on fine-leaved fescues under field conditions. Its ability to cause patch disease symptoms on other grasses is still under investigation.

Taxonomy. *G. incrustans* produces black, globose perithecia (Plate 19) with long cylindrical necks (Table 3.2a). Perithecia have been produced in culture (i.e., on axenic wheat roots) but have not been observed in the field. Asci are unitunicate, clavate (club-shaped), cylindrical, straight or slightly curved, eight-spored, and have a refractive ring at the apex (Plate 20). At maturity, ascospores are hyaline, fusoid (tapered at both ends), and have three to five septa (27) (Plate 21). Paraphyses are hyaline, septate, branched, and range 2-5 µm in diameter (26, 27).

Phialospores of *G. incrustans* are hyaline, subglobose to ovate, straight or slightly curved, and may have two rounded ends or one pointed and one rounded end (Table 3.3a) (Plate 21). The fungus persists on root surfaces as either brown runner hyphae (2-5 µm wide) or as slender, hyaline infection hyphae that penetrate root cortical tissue. Swollen hyphal cells typically fill root cortical cells and occur in large groups or as "crusts" on axenic root tissue. Hyphopodia are slightly lobed (6-14 µm long and 4-6 µm wide) and are borne singly or in groups on root surfaces. Hyphopodia are rarely observed in nature (26, 27).

In culture, mycelia are appressed, hyaline, and may turn gray to black with maturity (Plate 22). The black mycelial "crusts" that appear on the surfaces of older colonies (Plate 23) are characteristic of this fungus (27). The advancing edge of the mycelium often curls back towards the colony center. On half-strength potato dextrose agar, colonies advance at a rate of 7-14 mm$^{-24\,h}$ at 28-30 C (26).

Teleomorph and anamorph production. To successfully produce perithecia of *G. incrustans in vitro*, two compatible mating types should be placed on a minimal nutrient medium in the presence of host tissue (26). Landschoot and Jackson (27) produced perithecia by placing agar plugs from two compatible mating types on opposite sides of wheat stems that had been surface-sterilized with propylene oxide and embedded in Sach's agar (27, 29). Inoculated petri dishes were sealed with parafilm and incubated under diffuse light (i.e., from a west-facing window) for 8-10 wk. Another reliable method involves placing surface-disinfested wheat seedlings on water agar followed by inoculation with two compatible mating types (47).

Conidia of *G. incrustans* can be produced by placing plugs from recently obtained cultures on one-fifth strength potato dextrose agar (R. E. Wagner and H. T. Wilkinson, *personal communication*) or on dilute rabbit food agar. After storage in the dark at room temperature for 2-5 days, conidia may be observed on the agar surface. The author has found, however, that isolates of *G. incrustans* collected from the northeastern United States vary considerably in their ability to produce conidia using this method. Using an alternative technique developed by Speakman (47), conidia of *G. incrustans* may be observed after 3-5 wk. Observations are enhanced when cultures are inverted and viewed under low (10X) magnification.

Magnaporthe R. Krause & R. E. Webster

All species within the genus *Magnaporthe* are hyphopodiate parasites of the Gramineae (28). The most prominent feature of this genus is the presence of ascospores with darkly-pigmented inner cells and hyaline end cells. Both heterothallic and homothallic members have been identified. Currently, there are only four reported species of *Magnaporthe*: *M. salvinii* (Cattaneo) R. Krause & R. K. Webster; *M. grisea* (T. T. Hebert) Yaegashi & Udagawa; *M. rhizophila* Scott & Deacon; and *M. poae*. Only *M. poae* and *M. rhizophila* have *Phialophora* anamorphs (28).

Magnaporthe poae Landschoot & Jackson

M. poae was originally discovered in 1984 by Smiley and Craven Fowler (40). They found that the fungus (incorrectly reported as *P. graminicola*) was pathogenic to Kentucky bluegrass (*Poa pratensis* L.) at high (28 C) temperatures and determined that it was the causal agent of the disease summer patch. Landschoot and Jackson (28) later concluded that the causal fungus was a new species in the genus *Magnaporthe*. The pathogenicity of *M. poae* has also been demonstrated on fine-leaved fescues (25). *M. poae* and *M. grisea* (=*Pyricularia grisea* (Cooke) Sacc.) are the only heterothallic species within the genus and are the only species of *Magnaporthe* known to infect turfgrasses (28, 46).

Taxonomy. Perithecia of *M. poae* are black, globose, and have long, cylindrical necks (Table 3.2b) (Plate 43). To date, they have only been observed in culture (28, 41). Asci are clavate, eight-spored, straight to slightly curved, unitunicate, and have a refractive ring at the apex (Plate 44). Ascospores are fusoid and at maturity have three septa. The two terminal cells are hyaline and the intermediate cells are dark brown (Plate 45). Paraphyses are hyaline, septate, and are occasionally branched (28).

Phialospores of *M. poae* are hyaline and rounded at both ends (Table 3.3b) (Plate 48). Most are slightly curved, although some are straight. The runner hyphae of *M. poae* are sparse on host roots, brown, 2-5 µm wide, and frequently occur on stem bases (Plate 39). Infection hyphae are hyaline, slender, and originate from runner hyphae or hyphopodia. Hyphopodia are dark brown, slightly lobed, and globose (Plate 40), but are rarely observed on diseased tissue from the field. Incubation of stems and leaf sheaths under moist conditions for several days, however, occasionally yield hyphopodia (28, 46).

In culture, the mycelium of *M. poae* is appressed to the surface of the medium. It is hyaline and turns gray or olive-brown when mature (Plate 46). Thick, dark strands of mycelia radiate from the center of the colony. Hyphal strands often curl back towards the center of the petri dish (Plate 47) in a manner similar to *Gaeumannomyces* species. Maximum growth of *M. poae* occurs at 28-30 C, attaining a radius of 7-12 $mm^{-24\,h}$ (28).

Teleomorph and anamorph production. *M. poae* is a heterothallic species and, therefore, requires the successful union of two compatible mating types for the formation of perithecia (26, 28). Perithecia have been routinely produced by placing agar plugs from compatible mating types on a minimal nutrient agar in the presence of sterile grass stems, roots, or seeds. The fastest and most reliable method for producing perithecia and conidia of *M. poae* is similar to that described by Speakman (47). Isolates of *M. poae* appear to vary in their ability to form perithecia. It is, therefore, advisable to select isolates that produce abundant perithecia when attempting to identify unknown isolates.

Leptosphaeria Ces. & De Not., nom. cons.

The genus *Leptosphaeria* is characterized by the presence of pseudothecia, bitunicate asci, and brown ascospores. Several members of the genus are parasitic on the Gramineae (54). Smith (44) implicated *Ophiobolus herpotrichus* (Fr.:Fr.) Sacc. & Roum. as the causal agent of the spring dead spot disease of bermudagrass in Australia. Subsequent studies revealed that another fungus with shorter ascospores was the more common incitant of the disease (45). Walker and Smith (54) later re-examined Smith's collections and determined that both fungi belonged in the genus *Leptosphaeria*. *L. narmari* J. C. Walker & A. M. Sm. and *L. korrae* J. C. Walker & A. M. Sm. are now the accepted names of these pathogens. *L. korrae* is also the causal agent of the disease necrotic ring spot (61) and is one of the primary incitants of spring dead spot in the United States (6, 16).

Leptosphaeria korrae J. C. Walker & A. M. Sm.

Taxonomy. Most pseudothecia of *L. korrae* are flask-shaped and erumpent (Table 3.2b) (Plates 13-15). The necks are lined with periphyses and have a thickened ridge of cells at their base. The asci are bitunicate, cylindrical to clavate, eight-spored, and narrow to a foot-like support. Ascospores are filiform, slightly twisted in the ascus, pale brown, and contain up to 15 septa (Plate 16). The end cells of each spore are rounded and taper towards the base (43, 46, 54).

A conidial state has not yet been reported for *L. korrae*. Runner hyphae are brown, septate, and range from 2.5-5.0 µm in width. They often occur in strands on the surfaces of host roots, rhizomes, and stem bases and form flattened, dark sclerotia (Table 3.3b) that are 50-400 µm in diameter (6, 43, 46, 54).

On potato dextrose agar, mycelial growth is at first white but at maturity often turns dark gray to black. Some isolates, however, may remain white or light gray (Plates 17 and 18). The mycelium is aerial and has a felt-like appearance. Optimum radial growth is 4-5 mm$^{-24\,h}$ at 25 C (43, 46, 54).

Teleomorph production. Pseudothecia of *L. korrae* have been produced on leaf sheaths, stem bases, and roots of inoculated cereals and turfgrasses growing in sand or in sand/compost mixtures (6, 44, 61). Mature pseudothecia are usually produced in 6-8 wk. Crahay et al (6) reported pseudothecial formation on washed bermudagrass roots that had been placed on moistened gauze. When incubated at 22 C, pseudothecia were produced in 6 wk. Hammer and Chastagner

(22) also obtained mature pseudothecia by inoculating Scaldis hard fescue (*F. longifolia* Thuill.) seedlings grown on 2% water agar. They found that the optimal temperature and osmotic potential required for the formation of pseudothecia was 20 C and -0.06 MPa, respectively. Pseudothecia, however, have not been produced in culture without the presence of host tissue (46, 54).

Leptosphaeria narmari J. C. Walker & A. M. Sm.

Taxonomy. Pseudothecia of *L. narmari* develop within leaf sheaths or form superficially on stolons (46). They are black, flask-shaped, and occur singly or in clusters (Table 3.2b). The necks are lined with upwardly projecting periphyses and often have two thickened ridges of cells at the base. Asci are clavate (with a foot-shaped base), bitunicate, and eight-spored. Ascospores are biseriate, pale brown, elliptical to fusiform, and have three to seven septa (46, 54).

No conidial state of *L. narmari* has been found in the field or in culture. Runner hyphae on host tissue are brown, septate, and branched (2.5-5.0 µm wide). The hyphae are often fused into strands that form flattened, dark sclerotia that are 40-400 µm in diameter (43, 46, 54) (Table 3.3b).

On potato dextrose agar, the aerial mycelium of *L. narmari* changes from white to buff in color and gradually darkens as the colony matures. Optimum radial growth (4-5 mm$^{-24 h}$) occurs at 25 C (43, 46, 54).

Teleomorph production. Pseudothecia have been produced on both artificially inoculated and naturally infected plants (54). The *in vivo* techniques used to produce pseudothecia of *L. korrae* may also be used to produce pseudothecia of *L. narmari*. The teleomorph has not been produced in culture without the use of host tissue (54).

Ophiosphaerella Speg.

Walker (52) confirmed that *Ophiosphaerella* is a Pleosporaceous genus with a scolecospored conidial state and is, therefore, distinct from the genus *Ophiobolus*. It has been isolated from a number of plants in the Gramineae and Cyperaceae (49, 52).

Ophiosphaerella herpotricha (Fr.:Fr.) J. C. Walker
syns: *Ophiobolus herpotrichus* (Fr.:Fr.) Sacc. & Roum.; *Phaeosphaeria herpotricha* (Fr.:Fr.) L. Holm.; *Ophiobolus medusae* Ellis & Everh. f. *bromi* Brenckle; *Ophiobolus oryzae* Miyabe; and *Scolecosporiella sp.* (anamorph)

Until recently, *O. herpotricha* was considered to be a saprophyte or weak pathogen of the Gramineae (49, 52). In 1989, however, Tisserat et al (49) reported that *O. herpotricha* is a causal agent of spring dead spot of bermudagrass in Kansas. *O. herpotricha* has also been isolated from barnyardgrass (*Echinochloa crusgalli* (L.) P. Beauv.), bermudagrass, bromegrass (*Bromus inermis* Leyss.), corn (*Zea mays* L.), rice (*Oryza sativa* L.), *Vetiveria zizanioides* (L.) Nash, and zoysia (*Zoysia japonica* Steud.) (21, 49, 52).

Taxonomy. Pseudothecia of *O. herpotricha* are occasionally found on host tissue in the field (53). They are dark brown or black, spherical or flattened at the base, and have a neck that often protrudes through the leaf sheath of the host (Table 3.2b). The pseudothecium is surrounded by numerous dark brown, thick-walled hairs. Asci are bitunicate, cylindrical to club-shaped, and are surrounded by hyaline pseudoparaphyses. Ascospores are long, filiform, yellow to pale brown, eight-spored, and contain eight to 20 septa. The ascospores are parallel or loosely twisted within the ascus (49, 52, 53, 55).

Pycnidia are rarely produced in culture. Webster and Hudson (55), however, did produce the pycnidial state (*Scolecosporiella* sp.) of *O. herpotricha* on oat agar after an incubation period of 6-10 wk. When present, pycnidia resemble the ascocarp and are 200-550 µm in diameter. Conidia are cylindrical and contain five to six septa (Table 3.3b). The spores have a truncate base and taper to a bristle-like apex. Runner hyphae are dark brown, 3-7 µm wide, septate, and branched. Hyphae may produce intercalary hyphopodia on host tissue (53, 55).

   O. herpotricha produces a white, cottony mycelium on potato dextrose agar and malt agar. Cultures turn tan to brown in 3-7 days. Optimum growth ranges from 3.5-4 mm$^{-24\,h}$ at 20-25 C (49, 53, 55).

Teleomorph production. Tisserat et al (49) reported the formation of fertile pseudothecia on stolons and crowns of inoculated bermudagrass plants in the greenhouse. All attempts to initiate pseudothecia of *O. herpotricha* in culture have thus far been unsuccessful (49).

Phialophora Medlar. Conidial States

   Phialophora species are common inhabitants of grass roots (7, 32, 37, 38). According to Cole and Kendrick (5), the genus *Phialophora* represents an artificial grouping of unrelated species that produce conidia on phialides. Although considerable variation exists within the *Phialophora* conidial state, Walker (52) concluded that certain characteristics are interrelated (e.g., isolates that have slightly lobed hyphopodia typically grow slowly (4-6 mm$^{-24\,h}$)). To accurately identify species within this genus, multiple features must be examined. Schol-Schwarz (36) suggests comparing unknown isolates with herbaria specimens or culture collections, utilizing recognized authorities to verify cultures, and comparing multiple taxonomic characters to enhance the accuracy of species identification.

Phialophora graminicola (Deacon) J. Walker

    syn: *Phialophora radicicola* Cain var. *graminicola* Deacon

    P. graminicola has been intensively studied in Europe and in the United States with regard to its impact on cereal and grass hosts (40, 42, 53). The capacity of *P. graminicola* to protect roots against attack by *G. graminis* var. *tritici* and *G. graminis* var. *avenae* (see Landschoot et al, Chapter four) has also been documented (7, 8, 59). The pathogenicity of this fungus on turfgrasses, however, is still in question (26). *P. graminicola* has been isolated from annual ryegrass (*Lolium multiflorum* Lam.), bentgrass, fine-leaved fescue, Kentucky bluegrass, orchardgrass (*Dactylis glomerata* L.), perennial ryegrass (*L. perenne* L.), tall fescue (*F. arundinaceae* Schreb.), and timothy (*Phleum nodosum* L.) (7, 26, 52). Hornby et al (24) have suggested that *G. cylindrosporus* D. Hornby, D. Slope, R. Gutteridge & Sivanesan (Table 3.2a and 3.3a) is a possible teleomorph of *P. graminicola* (reported as *P. radicicola* var. *graminicola*).

    Taxonomy. Cultures are initially white but turn tan or gray as the colony matures. On potato dextrose agar, *P. graminicola* produces a felt of aerial mycelia. Phialides are hyaline to pale brown and are 5-20 μm long X 2-4 μm wide. They are produced singly on hyphae or in loose clusters at the ends of hyphal branches. Phialospores are hyaline, oblong, straight or slightly curved (5-11(14) μm long X 1.5-2.5 μm wide), and are rounded at the apex. They may be produced using the methods listed in this chapter for *G. incrustans*. Runner hyphae are brown, septate, and are 3-5 μm wide. Hyphopodia are simple (Plate 41) or slightly lobed and are numerous on grass roots in the field or on wheat coleoptiles in culture. Optimum radial growth is 4-6 mm$^{-24\,h}$ at 25 C (43, 52, 53).

Phialophora radicicola Cain

   syn: *Phialophora radicicola* var. *radiciola* sensu Deacon

The name *P. radicicola* has been used in the literature to describe several different fungi (3, 18, 32, 52). As a result, there is now a considerable degree of confusion surrounding its usage. Walker (52) re-examined the original collection of *P. radicicola* deposited in the University of Toronto's Herbarium, a subculture from the living type culture in the Centraal Bureau voor Schimmelcultures, and a culture of the original isolate from McKeen. Following a thorough examination of this material, Walker concluded that the original descriptions provided by Cain (3) and McKeen (32) and the current state of the type culture do not allow for valid comparisons between *P. radicicola* Cain and present-day isolates from the Gramineae. Thus, Walker (52) recommended that the name *P. radicicola* Cain should only be used to refer to the original type culture. If this concept is accepted, then all prior references to *P. radicicola* are in question. In particular, Freeman and Augustin's (18) report associating *P. radicicola* Cain with bermudagrass decline in Florida may be invalid.

   Taxonomy. On weak peptone malt-yeast extract medium, colonies of *P. radicicola* Cain are at first white but then turn gray with age. Phialides and phialospores are produced on this medium in about one week. Phialides are hyaline, 10-23 µm long X 3-4 µm wide, straight or slightly curved, and taper to a thin apex. They are usually found in groups at the end of branches or laterally at the ends of subterminal hyphal cells. A collarette is present at the tip of each phialide. Phialospores are produced at the phialide apex and are hyaline. Individual phialospores are curved, broadest near the base, and taper to a rounded apex (52, 53).

SUMMARY

Patch diseases of turfgrasses are caused by a closely related group of ERI fungi. These fungi produce a number of taxonomic characters that are useful in distinguishing species. The characteristics that are most helpful in identifying these fungi include the color, size, and shape of the ascocarps, asci, ascospores, conidia, and hyphopodia, as well as their growth in culture. A number of *in vitro* and *in vivo* techniques are available for producing these structures, although several weeks are usually required for their formation.

ACKNOWLEDGMENTS

Appreciation is extended to Dr. R. E. Wagner and Dr. H. T. Wilkinson for providing the unpublished information presented in this review.

LITERATURE CITED

1. Ainsworth, G. C., Sparrow, F. K., and Sussman, A. S., eds. 1973. The Fungi: An Advanced Treatise. Vol. IVA. Academic Press, New York.
2. Alexopoulos, C. J., and Mims, C. W. 1979. Introductory Mycology. 3rd ed. John Wiley & Sons, New York.
3. Cain, R. F. 1952. Studies of Fungi Imperfecti. I. *Phialophora*. Can. J. Bot. 30:338-343.
4. Chambers, S. C., and Flentje, N. T. 1967. Studies on variation with *Ophiobolus graminis*. Aust. J. Biol. Sci. 20:941-951.
5. Cole, G. T., and Kendrick B. 1973. Taxonomic studies of *Phialophora*. Mycologia 65:661-688.
6. Crahay, J. N., Dernoeden, P. H., and O'Neill, N. R. 1988. Growth and pathogenicity of *Leptosphaeria korrae* in bermudagrass. Plant Dis. 72:945-949.
7. Deacon, J. W. 1974. Further studies on *Phialophora radicicola* and *Gaeumannomyces graminis* on roots and

stem bases of grasses and cereals. Trans. Br. Mycol. Soc. 63:307-327.

8. Deacon, J. W. 1976. Biology of the *Gaeumannomyces graminis* Arx & Olivier/*Phialophora radicicola* Cain complex on roots of the Gramineae. EPPO Bull. 6:349-363.

9. Deacon, J. W., and Scott, D. B. 1983. *Phialophora zeicola* sp. nov., and its role in the root rot-stalk rot complex of maize. Trans Br. Mycol. Soc. 81:247-262.

10. Dernoeden, P. H., and O'Neill, N. R. 1983. Occurrence of Gaeumannomyces patch disease in Maryland and growth and pathogenicity of the causal agent. Plant Dis. 67:528-532.

11. Deverall, B. J., Wong, P. T. W., and McLeod, S. 1979. Failure to implicate antifungal substances in cross-protection of wheat against take-all. Trans Br. Mycol. Soc. 72:233-236.

12. Elliott, M. L. 1991. Determination of an etiological agent of bermudagrass decline. Phytopathology 81:1380-1384.

13. Elliot, M. L., Hagan, A. K., and Mullen, J. M. 1993. Association of *Gaeumannomyces graminis* var. *graminis* with a St. Augustinegrass root rot disease. Plant Dis. 77:206-209.

14. Elliott, M. L., and Landschoot, P. J. 1991. Fungi similar to *Gaeumannomyces* associated with root rot of turfgrasses in Florida. Plant Dis. 75:238-241.

15. Emmett, R. W., and Parbery, D. G. 1975. Appressoria. Annu. Rev. Phytopathol. 13:147-167.

16. Endo, R. M., Ohr, H. D., and Krausman, E. M. 1985. *Leptosphaeria korrae*, a cause of the spring dead spot disease of bermudagrass in California. Plant Dis. 69:235-237.

17. Farr, D. F., Bills, G. F., Chamuris, G. P., and Rossman, A. Y. 1989. Fungi on Plants and Plant Products in the United States. American Phytopathological Society, St. Paul, MN.

18. Freeman, T. E., and Augustin, B. J. 1986. Association of *Phialophora radicicola* Cain with declining bermudagrass in Florida. (Abstr.) Phytopathology 76:1057.

19. Gindrat, D. 1966. Etude de la formation des périthèces de *Gaeumannomyces graminis* (Sacc.) von Arx et Olivier en milieu artificiel et mise au point d'une technique d'immersion des cultures pour leur formation induite. Ber. Schweiz. Bot. Ges. 76:157-175.

20. Gindrat, D. 1968. Recherches sur la nutrition et le développement de *Gaeumannomyces graminis* (Sacc.) von Arx et Olivier, agent du piétin-échaudage des céréales. Schweiz. Landwirt. Forsch. 7:197-214.

21. Green, D., II, Fry, J., Pair, J., and Tisserat, N. 1992. Pathogenicity of fungi associated with a patch disease of zoysiagrass in Kansas. (Abstr.) Phytopathology 82:1123.

22. Hammer, W., and Chastagner, G. 1987. Factors affecting the production of pseudothecia of *Leptosphaeria korrae in vitro*. (Abstr.) Phytopathology 77:1239.

23. Holden, M., and Hornby, D. 1981. Methods of producing perithecia of *Gaeumannomyces graminis* and their application to related fungi. Trans. Br. Mycol. Soc. 77:107-118.

24. Hornby, D., Slope, D. B., Gutteridge, R. J., and Sivanesan, A. 1977. *Gaeumannomyces cylindrosporus*, a new ascomycete from cereal roots. Trans. Br. Mycol. Soc. 69:21-25.

25. Kemp, M. L., Clarke, B. B., and Funk, C. R. 1990. The susceptibility of fine fescues to isolates of *Magnaporthe poae* and *Gaeumannomyces incrustans*. (Abstr.) Phytopathology 80:978.

26. Landschoot, P. J. 1988. Taxonomy and pathogenicity of ectotrophic fungi with *Phialophora* anamorphs from turfgrasses. Ph.D. dissertation. University of Rhode Island, Kingston, Rhode Island.

27. Landschoot, P. J., and Jackson, N. 1989. *Gaeumannomyces incrustans* sp. nov., a root-infecting hyphopodiate

fungus from grass roots in the United States. Mycol. Res. 93:55-58.

28. Landschoot, P. J., and Jackson, N. 1989. *Magnaporthe poae* sp. nov., a hyphopodiate fungus with a *Phialophora* anamorph from grass roots in the United States. Mycol. Res. 93:59-62.

29. Luttrell, E. S. 1958. The perfect stage of *Helminthosporium turcicum*. Phytopathology 48:281-287.

30. Luttrell, E. S. 1973. Loculoascomycetes. Pages 135-219 in: The Fungi: an Advanced Treatise. Vol. IVA. G. C. Ainsworth, F. K. Sparrow, and A. S. Sussman, eds. Academic Press, New York.

31. McCarty, L. B., and Lucas, L. T. 1989. *Gaeumannomyces graminis* associated with spring dead spot of bermudagrass in the southeastern United States. Plant Dis. 73:659-661.

32. McKeen, W. E. 1952. *Phialophora radicicola* Cain, a corn rootrot pathogen. Can. J. Bot. 30:344-347.

33. Muller, E., and von Arx, J. A. 1973. Pyrenomycetes: Meliolales, Coronophorales, Sphaeriales. Pages 87-132 in: The Fungi: an Advanced Treatise. Vol. IVA. G. C. Ainsworth, F. K. Sparrow, and A. S. Sussman, eds. Academic Press, New York.

34. Nilsson, H. E. 1969. Studies of root and foot rot diseases of cereals and grasses. I. On resistance to *Ophiobolus graminis* Sacc. Lantbrukshögsk. Ann. 35:275-807.

35. Nilsson, H. E., and Smith, J. D. 1981. Take-all of grasses. Pages 433-488 in: Biology and Control of Take-all. M. J. C. Asher and P. J. Shipton, eds. Academic Press, New York.

36. Schol-Schwarz, M. B. 1970. Revision of the genus *Phialophora* (Moniliales). Persoonia 6:59-94.

37. Scott, P. R. 1970. *Phialophora radicicola*, an avirulent parasite of wheat and grass roots. Trans. Br. Mycol. Soc. 55:163-167.

38. Sivasithamparam, K. 1975. *Phialophora* and *Phialophora*-like fungi occurring in the root region of wheat. Aust. J. Bot. 23:193-212.

39. Sivasithamparam, K., and Parker, C. A. 1981. Physiology and nutrition in culture. Pages 125-150 in: Biology and Control of Take-all. M. J. C. Asher and P. J. Shipton, eds. Academic Press, New York.

40. Smiley, R. W., and Craven Fowler, M. 1984. *Leptosphaeria korrae* and *Phialophora graminicola* associated with Fusarium blight syndrome of *Poa pratensis* in New York. Plant Dis. 68:440-442.

41. Smiley, R. W., Dernoeden, P. H., and Clarke, B. B. 1992. Compendium of Turfgrass Diseases. 2nd ed. American Phytopathological Society, St. Paul, MN.

42. Smiley, R. W., Fowler, M. C., and Kane, R. T. 1985. Temperature and osmotic potential effects of *Phialophora graminicola* and other fungi associated with patch diseases of *Poa pratensis*. Phytopathology 75:1160-1167.

43. Smiley, R. W., Kane, R. T., and Craven-Fowler, M. 1985. Identification of *Gaeumannomyces*-like fungi associated with patch diseases of turfgrasses in North America. Pages 609-618 in: Proc. Int. Turfgrass Res. Conf., 5th. F. Lemaire, ed. INRA Publications and the International Turfgrass Society, Versailles, France.

44. Smith, A. M. 1965. *Ophiobolus herpotrichus*, a cause of spring dead spot in couch turf. Agric. Gaz. N. S. W. 76:753-758.

45. Smith, A. M. 1971. Spring dead spot of couch grass turf in New South Wales. J. Sports Turf Res. Inst. 47:54-59.

46. Smith, J. D., Jackson, N., and Woolhouse, A. R. 1989. Fungal Diseases of Amenity Turf Grasses. E. & F. N. Spon, London.

47. Speakman, J. B. 1982. A simple, reliable method of producing perithecia of *Gaeumannomyces graminis* var. *tritici* and its application to isolates of *Phialophora* spp. Trans. Br. Mycol. Soc. 79:350-353.

48. Speakman, J. B. 1984. Perithecia of *Gaeumannomyces graminis* var. *graminis* and *G. graminis* var. *tritici* in pure culture. Trans. Br. Mycol. Soc. 82:720-723.

49. Tisserat, N. A., Pair, J. C., and Nus, A. 1989. *Ophiosphaerella herpotricha*, a cause of spring dead spot of bermudagrass in Kansas. Plant Dis. 73:933-937.

50. Turner, E. M. 1940. *Ophiobolus graminis* Sacc. var. *avenae* var. n., as the cause of take all or whiteheads of oats in Wales. Trans. Br. Mycol. Soc. 24:269-281.

51. Walker, J. 1972. Type studies on *Gaeumannomyces graminis* and related fungi. Trans. Br. Mycol. Soc. 58:427-457.

52. Walker, J. 1980. *Gaeumannomyces, Linocarpon, Ophiobolus* and several other genera of scolecospored Ascomycetes and *Phialophora* conidial states, with a note on hyphopodia. Mycotaxon 11:1-129.

53. Walker, J. 1981. Taxonomy of take-all fungi and related genera and species. Pages 15-74 in: Biology and Control of Take-all. M. J. C. Asher and P. J. Shipton, eds. Academic Press, New York.

54. Walker, J., and Smith, A. M. 1972. *Leptosphaeria narmari* and *L. korrae* spp. nov., two long-spored pathogens of grasses in Australia. Trans. Br. Mycol. Soc. 58:459-466.

55. Webster, J., and Hudson, H. J. 1957. Graminicolous pyrenomycetes. VI. Conidia of *Ophiobolus herpotrichus, Leptosphaeria luctuosa, L. fuckelii, L. pontiformis* and *L. eustomoides*. Trans. Br. Mycol. Soc. 40:509-522.

56. Weste, G. 1970. Factors affecting vegetative growth and the production of perithecia in culture by *Ophiobolus graminis*. I. Variations in media and age of mycelium. Aust. J. Bot. 18:1-10.

57. Weste, G. 1970. Factors affecting vegetative growth and the production of perithecia in culture by *Ophiobolus graminis*. II. Variations in light and temperature. Aust. J. Bot. 18:11-28.

58. Weste, G., and Thrower, L. B. 1963. Production of perithecia and microconidia in culture by *Ophiobolus graminis*. Phytopathology 53:354.

59. Wong, P. T. W., and Siviour, T. R. 1979. Control of Ophiobolus patch in *Agrostis* turf using avirulent fungi and take-all suppressive soils in pot experiments. Ann. Appl. Biol. 92:191-197.

60. Wong, P. T. W., and Worrad, D. J. 1989. Preventative control of take-all patch of bentgrass turf using triazole fungicides and *Gaeumannomyces graminis* var. *graminis* following soil fumigation. Plant Prot. Q. 4:70-72.

61. Worf, G. L., Stewart, J. S., and Avenius, R. C. 1986. Necrotic ring spot disease of turfgrass in Wisconsin. Plant Dis. 70:453-458.

ECOLOGY AND EPIDEMIOLOGY OF ECTOTROPHIC ROOT-INFECTING FUNGI ASSOCIATED WITH PATCH DISEASES OF TURFGRASSES

Peter J. Landschoot
The Pennsylvania State University
University Park, PA 16802

and

Ann B. Gould and **Bruce B. Clarke**
Rutgers, The State University
New Brunswick, NJ 08903

INTRODUCTION

An intensively managed turf provides an ideal environment for the development of both foliar and root-infecting fungi. A typical stand of turf is composed of numerous plants of similar genotype in close proximity to one another. These plants are subject to a variety of stress-related conditions that increase their susceptibility to disease. Such stresses include regular mowing, heavy traffic, extremes in temperature and moisture, and the non target effects of pesticides (24). In addition, turf swards contain a surface layer of leaves, stems, and roots in various stages of decomposition. This layer provides a habitat and food source for various microorganisms, including pathogenic fungi.

73

Several ectotrophic root-infecting (ERI) fungi have been implicated as causal agents of turfgrass patch diseases such as take-all patch (*Gaeumannomyces graminis* (Sacc.) Arx & D. Olivier var. *avenae* (E. M. Turner) Dennis); summer patch (*Magnaporthe poae* Landschoot & Jackson); necrotic ring spot (*Leptosphaeria korrae* J. C. Walker & A. M. Sm.); spring dead spot (*L. korrae, L. narmari* J. C. Walker & A. M. Sm., *G. graminis* (Sacc.) Arx & D. Olivier var. *graminis,* and *Ophiosphaerella herpotricha* (Fr.:Fr.) J. C. Walker); and bermudagrass decline (*G. graminis* var. *graminis*). These fungi can exploit the turfgrass ecosystem and are well equipped to persist on root surfaces or in cortical cells for long periods of time without causing visible injury. When conditions favor disease development, however, they may invade the vascular tissue, resulting in root dysfunction and eventual death of the turf.

A thorough understanding of the factors that affect the growth and development of the ERI fungi is crucial to the effective management of turfgrass patch diseases. Unfortunately, since most of these diseases and their causal agents have only recently been identified on turfgrass hosts, the ecology and epidemiology of the ERI fungi is poorly understood. This chapter is a review of the literature pertaining to the colonization, saprophytic survival, and dissemination of the ERI fungi. In addition, the influence of environmental factors on disease development and the interactions between the ERI fungi and other soilborne microorganisms is discussed.

COLONIZATION

One characteristic common to all ERI fungi is their "ectotrophic growth habit." Garrett (33) used this term to describe the continuous, external growth of fungal hyphae over root surfaces that can occur prior to penetration of root cortical cells. Unlike most ERI fungi, the ectotrophic growth habit of *G. graminis* (Sacc.) Arx. & D. Olivier var. *tritici* J. Walker, the causal agent of take-all of wheat, and *G. graminis* var. *avenae,* the causal agent of take-all of bentgrass (*Agrostis* spp.) and small grains, has been extensively studied (33, 34, 55, 56, 69, 87, 94). In his book "Pathogenic Root-Infecting Fungi," Garrett (34) provides an in-depth analysis of the ectotrophic growth habit of *G. graminis* (reported as *Ophiobolus graminis* (Sacc.) Sacc. in Roum. & Sacc.). Although it is likely that some differences exist, the mechanisms described by Garrett and others may serve as a model for many of the ERI fungi that attack turf.

When wheat (*Triticum aestivum* L.) roots are inoculated with *G. graminis* (reported as *O. graminis*), ectotrophic hyphae are at first hyaline to yellow-brown (94). Within 24-30 h, hyphal strands grow together and fuse laterally to form mature, dark brown "macro-" or "runner hyphae." Ectotrophic runner hyphae usually grow longitudinally over root surfaces (69). The hyphae may then anastomose and either grow as a mycelial mat, a mass of pseudoparenchymatous tissue, or develop into sclerotia-like bodies (69). On turfgrasses, the ectotrophic mycelia may infest roots, crowns, stolons, rhizomes, stem bases, and leaf sheaths (26, 56, 72, 76, 101). The mycelia may also differentiate into specialized attachment and infection structures called hyphopodia (Plate 60) from which infection hyphae can penetrate cortical tissue (69).

Ectotrophic growth (34) and patch disease development (19, 51, 73, 75) have been shown to increase following a decline in plant vigor. There is little direct evidence, however, to support the hypothesis that an increase in ectotrophic growth results in an increase in the severity of patch disease. Garrett (34) speculates that an interactive relationship exists between the defense mechanisms of the host and ectotrophic growth. In vigorously growing plants, the author suggests that the host restricts colonization by the pathogen to an ectotrophic growth habit on the root. If host defense mechanisms weaken, however, infection may proceed with minimal obstruction. The author also suggests that the ectotrophic network may initiate multiple infection foci to overcome the host's defenses. A complete failure of these defenses, however, may not necessarily be advantageous to the ERI fungi since less specialized root pathogens may also gain ingress and compete for available organic substrates. Thus, at an intermediate level, the host's defense mechanisms may permit infection by ectotrophic fungi while limiting invasion by non specialized pathogens and saprophytes.

SAPROPHYTIC EXISTENCE

To understand how patch diseases develop in newly established turf, it is important to understand the manner in which these fungi are disseminated and survive saprophytically. Can the ERI fungi, for example, survive in topsoil or on equipment used for lawn establishment? Are they transported in sod or in seed? The answers to these questions could help explain why patch diseases often occur on sites with no previous history of patch disease.

In annual crops, root diseases can often be managed through crop rotation or by leaving the soil fallow between plantings. This practice reduces the food reserves on which pathogens exist. Roots of perennial turfgrasses such as Kentucky bluegrass (*Poa pratensis* L.), however, may grow throughout the year and are continually replaced (81). Consequently, the ERI fungi that infect turfgrass roots are often associated with living tissue for long periods. Saprophytic existence, therefore, may not be as critical to the survival of these fungi as it is for root pathogens on annual crops. Even so, ERI fungi must be able to persist in a saprophytic state if they are to survive in soils devoid of living plants. Such situations may occur if a turf is killed by insects or disease, or is destroyed by herbicides during renovation and is not immediately reseeded or sodded.

Survival of the ERI fungi in the absence of living turf roots is not well characterized. This is understandable since many of these fungi have only recently been discovered on turfgrasses. The saprophytic survival of *G. graminis* var. *tritici* has been extensively reviewed by Garrett (34) (reported as *O.*

graminis) and Shipton (67). This information will be utilized in the following discussion to examine the possible strategies for saprophytic survival among the ERI fungi that parasitize turf.

Saprophytic Survival in Soil

As a rule, pathogenic root-infecting fungi may survive in the absence of living host tissue as dormant resting structures or as saprophytes in dead organic substrates (34). *G. graminis*, and probably most other ERI fungi, rarely produce dormant resting structures (e.g., fruiting bodies, spores, chlamydospores, or sclerotia) that function as important survival structures in the soil (34, 67, 85). Although runner hyphae and pseudo-parenchymatous structures may function in this role, Hornby (40) states that the runner hyphae of *G. graminis* do not appear to be important survival structures in soil because of their low potential to infect seedlings. Most available evidence indicates that *G. graminis* survives in the soil primarily in dead plant tissues previously colonized through parasitism. The majority of specialized root-infecting pathogens such as *G. graminis* compete with non specialized pathogens and saprophytes more effectively for organic substrates when they are already in possession of the tissue before the plant dies (34). Whereas there have been relatively few cases of direct isolation from organic debris (28, 87), *G. graminis* has often been detected using trap hosts such as wheat (38). Hornby (38) used this technique to assess the infectivity of organic debris infested with *G. graminis* (reported as *O. graminis*). The author found that a coarse organic fraction exceeding 420 μm was more infective than finer debris or whole soil.

The potential for an ERI fungus to colonize an organic substrate without prior parasitism of the host tissue can be measured by its "competitive saprophytic ability." This term has been defined as the physiological characteristics necessary to enable a fungus to compete successfully against other soil microflora for colonization of dead organic substrates (33, 34).

To examine the competitive saprophytic abilities of various soilborne fungi, the Cambridge Method was developed (9, 10, 34, 49). In this procedure, a pure culture of a root-infecting fungus is mixed with sterile wheat straw and then diluted with increasing quantities of non sterile soil. Subsequent colonization of the wheat straw by the root-infecting fungus is then assessed. Using this method, Butler (9) found that the competitive saprophytic ability of *G. graminis* (reported as *O. graminis*) was reduced by 54% with the addition of 2% non sterile soil. At the same soil dilution, the competitive saprophytic abilities of *Fusarium culmorum* (Wm. G. Sm.) Sacc. and *Curvularia ramosa* (Bainier) Boedjin were only reduced by 13% and 4%, respectively. The author concluded that *G. graminis*, when compared to other soil microflora, is a weak competitor for organic substrates. Utilizing a modification of the Cambridge Method called the Agar Plate Method, Rao (60) reported similar results. Based on these studies, Garrett (34) concluded that saprophytic colonization without prior parasitism appears to be of minor importance for *G. graminis*. This may also hold true for the ERI fungi that cause patch diseases on turf.

Influence of Soil factors on Saprophytic Survival

Although *G. graminis* can withstand a wide range of soil conditions, it can only do so for relatively short periods. Under conditions that favor microbial activity (e.g., medium to high temperatures, adequate soil moisture, and good soil aeration), *G. graminis* and other soil microorganisms quickly exhaust available organic substrates (30, 34). When this occurs, saprophytic survival and longevity of *G. graminis* is reduced due to its poor competitive ability and lack of survival structures. Conversely, saprophytic survival of the fungus increases when conditions are likely to restrict microbial activity in soil (i.e., in air-dried or saturated soils at low temperatures).

Fellows (27) found that *G. graminis* in naturally-infested soils could tolerate exposure to temperatures as high as 60-71 C for 1 h on each of several days. MacNish (50) reported that *G. graminis* remained viable for 45 wk under conditions that were either dry (-25.0 to -98.0 MPa soil matrix potential) and cool (15 C) or moist (-0.4 to -0.7 MPa) and cool (15 C). A significant reduction in viability occurred, however, under dry (-98.0 MPa) and hot (35 C) conditions or in a wet (-0.01 to -0.02 MPa) and cool (15 C) environment. Furthermore, viability of *G. graminis* was eliminated within 4 wk in a wet (-0.01 to -0.02 MPa) and hot (35 C) environment.

Another factor that strongly influences the survival of *G. graminis* in soil is the availability of nitrogen (67). In two studies, Garrett (30, 32) found that *G. graminis* (reported as *O. graminis*) maintained viability for a longer period in soils amended with nitrogen than in unamended, nitrogen deficient soils. Garrett (34) concluded that the addition of nitrogen to soils deficient in nitrogen is necessary for the fungus to assimilate carbohydrate reserves present within organic substrates.

In summary, *G. graminis* can tolerate and persist under a wide range of soil conditions. Compared to other soilborne pathogens, however, its ability to compete for organic substrate is limited; hence, in the absence of a living host, the fungus only remains viable in soil for relatively short periods of time. Although much is known about the saprophytic survival and longevity of *G. graminis*, additional studies need to be conducted with other ERI fungi to better understand the survival of these organisms in the soil environment.

DISSEMINATION

G. graminis can spread from plant to plant whenever healthy tissues (i.e., roots, stolons, and leaf sheaths) come into direct contact with infected roots or colonized organic debris (69). The fungus is also able to grow short distances through the soil from colonized substrates to susceptible host tissue (8,

59, 69). Although *G. graminis* and other ERI fungi produce ascospores (Plates 64 and 65), their role in the dissemination of these organisms is difficult to assess (40).

In 1933, Samuel and Garrett (64) found that ascospores of *G. graminis* (reported as *O. graminis*) are forcibly ejected into the air from perithecia. Although Garrett (31) and Brooks (7) were able to infect wheat roots with suspensions of ascospores in sterilized soil or sand, they were unable to infect roots in non sterile field soil. Brooks (6, 7), however, determined that ascospores of *G. graminis* (reported as *O. graminis*) could infect wheat roots grown on the surface of moist, unsterilized soil if they were placed on the proximal regions of seminal roots as the roots penetrated the soil. The author speculated that ascospores were able to infect roots if present on the root surface prior to colonization by other rhizosphere microflora (7). As a result of this research, Garrett (34) concluded that the long distance transport of ascospores to wheat fields could explain the occurrence of take-all in the newly drained regions of the Noord Oost Polder, a region in the Netherlands with no prior history of the disease.

The importance of ascospores in the dissemination of the ERI fungi that parasitize turf has not been assessed. The occurrence of take-all patch on recently fumigated golf greens seeded to bentgrass, however, could be explained by the air-borne dissemination of ascospores. Assuming that ascospores of *G. graminis* var. *graminis*, *L. korrae*, or *O. herpotricha* are ejected into the air and have the ability to infect turfgrasses, this phenomenon could also account for the spread of bermudagrass decline, necrotic ring spot, and spring dead spot in geographically isolated areas. Since fruiting bodies of *M. poae* have not been found in nature, ascospores are not presently considered to be a means of dissemination for this pathogen.

The most likely means of dissemination for the ERI fungi that infect turf appears to be through movement of infested root, crown, or stem tissue during cultivation. Pair et al (57) demonstrated that spring dead spot could be spread by the transfer of diseased turf cores into areas with no previous history of the disease. Fungal inoculum may be similarly transferred by core aerators, dethatching machines, or even on the spikes of golf shoes. Smith et al (79) also speculated that patch diseases can spread from one geographical location to another through the transport of infected sod. Although many pathogenic fungi are disseminated through infected seed, this method of dissemination has not been demonstrated for the ERI fungi that parasitize turf.

INFLUENCE OF ENVIRONMENTAL FACTORS ON DISEASE DEVELOPMENT

Take-all Patch

Moisture

The development of take-all in cereals and bentgrass is most severe during periods of cool, wet weather (12, 36, 78). This may be largely explained by the high water potential requirement of the pathogen, *G. graminis* (12, 15). Cook et al (15) measured the growth response of *G. graminis* (reported as *O. graminis*) over a range of osmotic and matric potentials at 20 C. In agar media amended with various salts, fungal growth was optimal between -0.12 and -0.15 MPa and decreased in a linear fashion with a corresponding reduction in osmotic potential. Growth of *G. graminis* was reduced by 50% at -2.0 MPa and ceased at -4.5 to -5.0 MPa. The authors reported similar results at various matric potentials in soil.

The factors that influence the growth of *G. graminis* in response to various water potentials include the variety of the fungus and the ambient temperature. Cook and Christen (13) assessed the influence of temperature on the water potential requirements of *G. graminis* var. *tritici* in culture. At 20 and 25 C, the greatest growth response occurred at osmotic potentials between -0.5 and -0.8 MPa. At 30 C, however, slightly lower osmotic potentials (-0.8 to -1.2 MPa) were required for optimal growth.

In a similar study, Wong (98) assessed the effect of osmotic potential and temperature on the growth of three varieties of *G. graminis*: *G. graminis* var. *tritici*; *G. graminis* var. *avenae*; and *G. graminis* var. *graminis*. At 20 C, the growth response for each fungus was optimal at -0.12 MPa and decreased with a reduction in osmotic potential. The osmotic potential that limited growth, however, differed for each fungus. Most isolates of *G. graminis* var. *avenae* tested did not grow at osmotic potentials lower than -5.0 MPa. The lower limit for *G. graminis* var. *tritici*, however, was -6.0 MPa. Finally, three out of ten isolates of *G. graminis* var. *graminis* tested were capable of limited growth at -8.0 MPa. The growth response of these fungi to various osmotic potentials was temperature dependent in a manner similar to that reported by Cook and Christen (13). At 30 C, optimal growth occurred at -1.0 to -1.5 MPa for *G. graminis* var. *tritici* and *G. graminis* var. *avenae*. The osmotic potential required for the optimum growth of *G. graminis* var. *graminis*, however, did not change until the temperature reached 35 C. This observation may represent a mechanism whereby *G. graminis* can adapt to heat and moisture stress by lowering its water potential requirements for growth (13, 98).

    G. graminis requires a higher water potential for growth than either the host or many other soilborne pathogens (12, 14, 58). Since the fungus colonizes wheat root and stem tissue in the tillage layer, it is particularly sensitive to moisture fluctuations in the top 25 cm of soil (12). For severe outbreaks of take-all to occur, high moisture levels in the surface layer of soil must be maintained. This may explain the infrequent occurrence of take-all in the dryland wheat-growing regions of the northwestern United States, where soil water potential in the tillage layer may be as low as -5.0 MPa by early June (15). This is far below the water potential required for optimal growth of *G. graminis* (15) and for many isolates may prevent growth completely.

    Although Wong (98) studied the water potential requirements of *G. graminis* var. *avenae* in culture, the influence of water potential on take-all in turfgrass stands has not been assessed. Take-all is generally more severe in areas that receive high annual precipitation (12, 74); however, the disease has also been observed in the late summer months following drought stress (41, 56). In such cases, turf plants may become infected in the spring when conditions are conducive to the growth of the fungus (12). As the infected roots become stressed by heat and drought, the plants may succumb to the disease and symptoms appear (Plate 57). Drought stress, however, may also reduce the activity of organisms normally antagonistic to *G. graminis*, thus enhancing disease development in the field (12).

Temperature

In culture, the growth of *G. graminis* var. *tritici* is optimal at 25 C (68). Similarly, the growth rate of *G. graminis* var. *avenae* (the causal agent of take-all patch) is greatest at 25 C and is completely inhibited at 35 C (22). In the field, however, take-all of wheat is most severe at 5-15 C (12, 55). Henry (37) reported that the temperature optimal for the growth of *G. graminis* (reported as *O. graminis*) *in vitro* was similar to the temperature required for optimum disease development only if the soil was autoclaved. Both Cook (12) and Henry (37) postulated that the disease may develop under cooler temperatures in the field because the activity of microbial antagonists is reduced. Once plants become infected at cool temperatures, however, the disease may continue to develop at higher temperatures (12).

Soil Edaphic Factors

Soil conditions optimal for the development of take-all patch in the field include a neutral to alkaline pH, high sand content, low fertility, and good soil aeration (74). Although *G. graminis* can grow over a wide pH range, growth in culture is optimal at a pH of 6 to 7 (68). In the field, take-all is most severe in soils where the pH exceeds 7.0 in the upper 2.5 cm of the soil profile (78). It has been suggested that, in such soils, populations of microorganisms antagonistic to *G.*

graminis are reduced (71). Reis et al (61) found that the severity of take-all in soils deficient in phosphorus, zinc, or copper was reduced following the application of the deficient element. In Chapter six, Dernoeden further discusses the role of soil acidification and fertility in patch disease control.

Although take-all of wheat and bentgrass is intensified in soils with good aeration and high sand content (29, 74), the disease can also develop on poorly drained and compacted sites. Under anaerobic conditions, plant vigor, host defense systems, and the activity of microbial antagonists may be greatly reduced (56). It would appear, therefore, that *G. graminis* can adapt to a habitat in which the infection process may proceed at an elevated rate with minimal interference from host defenses or microbial antagonism.

Summer Patch

Moisture

Prior to 1984, summer patch remained an unidentified component of the Fusarium blight syndrome (72, 74). Summer patch typically occurs on closely mowed turf from June through September (79) during periods of hot, rainy weather (44, 74). Drought stress was at first thought to be a major factor predisposing turf to the Fusarium blight syndrome (3, 4, 17, 25). Smiley (70), however, noted that this syndrome was most often associated with excessive moisture or alternating periods of rainfall and drought. Such conflicting observations were made before the causal relationship between *M. poae* and summer patch was established. As a result, the effect of environmental factors on summer patch development prior to 1984 is difficult to assess. To date, this disease has only been reported in North America (Fig. 2.2).

In 1985, Smiley et al (75) measured the growth response of *M. poae* (reported as *Phialophora graminicola* (Deacon) J. Walker) in a minimal salts medium adjusted to a range of osmotic potentials with potassium chloride. At 20 C, maximum growth of the fungus occurred at -0.1 MPa, the highest osmotic potential used in the experiment. Growth of *M. poae* was reduced by 50% at -0.6 MPa and was completely inhibited at -2.2 MPa.

In a series of three companion studies, Kackley et al (43, 44, 45) investigated the role of drought stress in summer patch development. *In vitro* (43), the interactive effects of temperature and osmotic potential on the growth response of *M. poae* were assessed. Temperature, followed by osmotic potential, accounted for the greatest amount of variation in fungal growth. The optimum temperature for growth was 30 C. At 25 and 30 C, growth of *M. poae* was greatest at -0.12 MPa and decreased with decreasing osmotic potential. At 35 C, however, the growth rate of the fungus was greater at slightly lower osmotic potentials (-1.47 to -1.08 MPa). The growth response of this fungus to various osmotic potentials is, therefore, temperature dependent in a manner similar to that reported for *G. graminis* var. *tritici* (13).

In growth chamber studies, Kackley et al (45) reported that summer patch on annual bluegrass (*P. annua* L.) and Kentucky bluegrass was generally more severe at higher soil matric potentials. At temperatures supraoptimal for these cool-season turfgrasses (25 and 30 C), the most severe symptom development occurred on non moisture stressed (-0.05 MPa) Kentucky bluegrass turf and on mildly stressed (-0.40 to -0.80 MPa) annual bluegrass turf. At a temperature closer to the optimum for sustained root growth of these grasses (20 C), overall disease severity was less than that reported for higher temperatures. In addition, summer patch was more severe on Kentucky bluegrass plants exposed to mild water stress (-0.40 MPa) at this temperature than on plants that were not drought stressed (-0.05 MPa).

Finally, Kackley et al (44) assessed the impact of drought stress (< -0.05 MPa soil matric potential) or non moisture stress treatments on mature Kentucky bluegrass in field plots inoculated with *M. poae*. At the sites evaluated, summer patch was more severe in plots that were not drought stressed. Based on data from all three studies, Kackley et al (44) concluded that temperature and water potential, but not drought stress, are the key predisposing factors to summer patch development.

Temperature

As mentioned above, temperature plays an important role in the development of summer patch (44, 45). Smiley et al (75) reported that maximum growth of *M. poae* (reported as *P. graminicola*) *in vitro* occurred at 27-31 C. Similarly, Landschoot and Jackson (46) found that most isolates of the fungus grew best in culture at 30 C, whereas the fungus did not grow at 10 or 40 C. Kackley et al (43) reported that the temperature optimal for the growth of *M. poae in vitro* and for pathogenicity *in vivo* was 30 C.

Smiley et al (75) assessed the effect of temperature on ectotrophic growth, pathogenicity, and spread of *M. poae* (reported as *P. graminicola*) *in vivo*. At 24 and 29 C, the extent of ectotrophic growth of the fungus along rhizomes buried in non sterile soil progressed at the rate of 2 cm^{-wk}. At 14 C, however, no ectotrophic growth was observed. In pot studies, *M. poae* colonized the roots of Merion Kentucky bluegrass at 14, 21, and 29 C, but visible disease symptoms were evident only at the highest temperature. When plants that had been grown at 14 or 21 C were further incubated at 29 C, however, they developed symptoms of the disease within 2 wk. Finally, the spread of summer patch symptoms in trays of sod inoculated with the fungus progressed at the rate of 3 cm^{-wk} at 21 C, whereas little disease expression was evident at 14 C.

Although temperature affects the growth of *M. poae*, it also affects root growth and the host's susceptibility to disease. Maximum root development in Kentucky bluegrass occurs when the soil temperature is between 10 and 18.5 C (5); at 26 C, root growth is considerably impaired (102). Kackley et al (44) speculated that under the warm (25-30 C), moist soil conditions optimal for the growth of *M. poae*, root development is impaired and host plants become more susceptible to disease. In cooler (less than 25 C), moist soils, conditions favor the host and disease development is minimal. At these temperatures, summer patch develops only when turf becomes stressed by drought. Smiley et al (75) stated that colonization of roots by the pathogen under conditions of optimal soil moisture invokes a level of vascular dysfunction that is not lethal to the plant until stress is imposed. Development of summer patch, therefore, appears to be dependent on the rate of pathogen growth, host root development, and stress (75).

Soil Edaphic Factors

Summer patch is observed most often on poorly drained, compacted soils (74). Little information is available concerning the effect of pH and soil fertility on summer patch development. A discussion of the role of mowing, soil acidification, and nitrogen fertility in summer patch control can be found in Chapter six.

Necrotic Ring Spot

Necrotic ring spot is a widespread disease of cool-season turfgrasses. Although the disease occurs throughout the growing season, symptoms frequently appear under the cool, moist conditions that favor the growth of the incitant, *L. korrae* (74, 79). The disease is most prevalent on two to eight year old lawns established from sod; however, seeded sites may also be affected (11, 101).

Moisture

Necrotic ring spot has been reported on both irrigated and non irrigated turf in New York State (75) and frequently appears under conditions of drought stress in Pennsylvania (P. L. Landschoot, *personal observation*). *L. korrae* grows optimally in culture at an osmotic potential of -0.1 MPa at 20 C (75). Although the optimum growth rate of *L. korrae*, *G. graminis* var. *avenae*, and *M. poae* occurs at a similar osmotic potential (75), *L. korrae* is capable of growing under much drier conditions than the other fungi. The growth response of *L. korrae* is reduced by 50% at -2.5 MPa and is not completely inhibited until -7.0 MPa (75). The growth of *G. graminis* var. *avenae* and *M. poae* (reported as *P. graminicola*), however, is completely inhibited at -4.5 (15) and -2.2 MPa (75), respectively.

Since necrotic ring spot occurs over a wider soil moisture range than either take-all patch or summer patch, drought stress may play a more important role in the development of this disease (74). In addition, *L. korrae* appears to be more tolerant of soil moisture extremes than its cool-season hosts (5, 74, 75). This may represent a competitive advantage for the pathogen, particularly during periods of high temperature and drought stress.

Temperature

In culture, the growth of *L. korrae* is optimal at 25 C (18, 75, 86). Fungal growth is greatly reduced at or below 10 C and is completely inhibited at 30-32 C (18, 75). Little is known, however, about the development of *L. korrae* in the field at different soil temperatures.

Worf et al (101) assessed the influence of temperature and humidity on necrotic ring spot development in pot studies. In inoculated annual bluegrass sod, disease development was more severe at 24 C than at 16 C. Similarly, symptoms were more apparent in inoculated Kentucky bluegrass at 28 C than at 20 C. Necrotic ring spot did develop at 20 C; however, exposure to high humidity was necessary to induce symptom expression at this temperature. The authors concluded that temperature was not the sole determinant for symptom expression and speculated that other factors (i.e., those affecting fungal growth) may play a more important role in disease development. To date, the impact of such factors on symptom severity and disease development has not been assessed.

Spring Dead Spot

In the United States, spring dead spot may occur wherever bermudagrass (*Cynodon dactylon* (L.) Pers.) is subject to winter dormancy and cold temperatures (Fig. 2.3) (48, 79). The disease is most prevalent during cool, moist weather in the spring and fall (18, 48). To date, *L. korrae, L. narmari, O. herpotricha*, and *G. graminis* var. *graminis* have been associated with this disease. *L. narmari*, however, has only been confirmed as a pathogen of bermudagrass in Australia and New Zealand (42, 86).

Moisture

Of the reported incitants of spring dead spot, the impact of moisture on fungal growth has only been assessed for *L. korrae* (75) and *G. graminis* var. *graminis* (98). The moisture requirements of *L. korrae*, also the incitant of necrotic ring spot, have been previously discussed in this chapter. The growth of *G. graminis* var. *graminis* in culture is optimal at -0.12 MPa and is inhibited at -8.0 MPa at 20 C (98).

Temperature

More information exists regarding the effect of temperature on fungal growth and spring dead spot development than any other environmental parameter. In culture, the growth of *L. korrae, L. narmari*, and *G. graminis* var. *graminis* is optimal at 25 C (18, 75, 86, 96). Temperatures greater than 30 C and less than 10 C, however, inhibit the growth of these fungi (18, 75, 86, 96). Optimum growth of *O. herpotricha* occurs between 20 and 25 C and is restricted at temperatures above 30 C and below 10 C (82). Although the temperature optimal for the growth of these pathogens is 25 C, the severity of spring dead spot in the field is typically associated with colder temperatures (47, 48, 77).

In pot studies, Endo et al (26) evaluated the effect of temperature on the severity of spring dead spot in Tifgreen bermudagrass inoculated with *L. korrae*. At 13 C, the growth of the bermudagrass host was poor and extensive decay was evident on the roots, buds, stolons, and lower stem bases. In contrast, the host plants appeared vigorous at 24 C and symptom development was minimal. Since the bermudagrass

89

in this experiment had not become dormant, the authors concluded that low temperature may be more important than host dormancy in the development of spring dead spot in the field.

In growth chamber studies, Crahay et al (18) reported findings similar to Endo et al (26) using Tufcote bermudagrass inoculated with *L. korrae*. Although *L. korrae* colonized roots at temperatures between 15 and 30 C, plant mortality only occurred between 15 and 20 C. Root length and tillering of both the inoculated and uninoculated plants were also reduced at the lower temperatures. At temperatures that are suboptimal (10-20 C) for the development of warm-season roots (5), the authors speculated that the capacity of bermudagrass roots to resist fungal infection is diminished.

Tisserat et al (82) inoculated Arizona common bermudagrass in the greenhouse with isolates of *O. herpotricha*, *G. graminis* var. *graminis*, and *L. korrae*. Although infection by all three fungi caused significant root discoloration and decreased root dry weight compared to controls, *O. herpotricha* was the most virulent pathogen tested. In plants infected with *O. herpotricha* or *L. korrae*, root weight was reduced at both 15 and 25 C. At either temperature, however, plant mortality was not induced by infection. The authors concluded that *O. herpotricha* and *L. korrae* may infect bermudagrass whenever soil temperatures are between 10 and 25 C, but that low temperatures or dormancy are required for the development of foliar symptoms in the field.

McCarty et al (51), reported findings similar to Tisserat et al (82) using Tifway bermudagrass infected with *G. graminis* var. *graminis*. In this growth chamber study, when compared to uninfected turf, regrowth of infected turf was reduced by 95% following exposure to -5 C. Furthermore, foliar injury caused by *G. graminis* var. *graminis* was enhanced as low temperature stress increased.

Bermudagrass Decline

Bermudagrass decline is a newly described root disease of highly managed bermudagrass turf (23, 74). *G. graminis* var. *graminis*, a fungus recently associated with spring dead spot of bermudagrass in North Carolina (52, 53), has recently been implicated as the causal agent of this disease (23). Unlike spring dead spot, which develops under cool, moist conditions, symptoms of bermudagrass decline occur during hot, humid, and wet weather (23, 74).

Little is known about the influence of moisture and temperature on the growth of *G. graminis* var. *graminis* in culture or on the development of bermudagrass decline in the field. *G. graminis* var. *graminis* grows well in culture at -0.12 MPa (98) and 25 C (96). The fungus, however, can tolerate conditions that are much drier (-8.0 MPa) (98) and hotter (30 C) (96) than those required for optimal growth *in vitro*. Since *G. graminis* var. *graminis* is adapted to a wide range of temperatures and osmotic potentials (96, 98), the type of symptom expressed in bermudagrass turf may be dependent upon the environment. For instance, where bermudagrass is subject to cold temperatures, symptoms of spring dead spot, induced by *G. graminis* var. *graminis*, are expressed in roughly circular patches of dead turf 5 cm to 1 m in diameter (79) (Plates 1-3). In warmer climates such as Florida, however, *G. graminis* var. *graminis* is associated with larger areas of blighted, chlorotic turf that are attributed to bermudagrass decline (Plates 66 and 67). This theory has yet to be verified since Koch's postulates have not been confirmed in the field for isolates of *G. graminis* var. *graminis* associated with either bermudagrass decline or spring dead spot (23).

RELATIONSHIPS WITH OTHER SOILBORNE ORGANISMS

The relationship of the ERI fungi to other soilborne microorganisms has generated considerable interest among researchers in recent years. This is undoubtedly the result of successful work conducted on the biological suppression of take-all in wheat. Since this subject has been extensively reviewed by Cook and Rovira (16), Deacon (21), Rovira and Wildermuth (63), Weller (90), and Wong (97), only a brief discussion is presented here.

Take-all Decline

The biological suppression of take-all of small grains was recognized as early as the 1930s (30, 37, 65). During this period, a phenomenon known as take-all decline was discovered (35). Take-all decline has been defined as the decrease in disease that often follows a severe outbreak in a continuously cropped system (39). Although a number of theories have been proposed to explain the mechanisms involved in take-all decline, a thorough discussion of these mechanisms is beyond the scope of this chapter. For an in-depth review of the subject, refer to Hornby (39), Rovira and Wildermuth (63), and Walker (84).

In take-all suppressive soils, suppression of the pathogen *G. graminis* var. *tritici* may be classified as either "general" or "specific" (16, 63). General suppression is related to the total microbial activity of the soil. This activity is not destroyed by moist heat (70 C) and is not transferrable to other soils. On the other hand, specific suppression is associated with a cereal monoculture, is sensitive to moist heat at 60 C, and is transferrable to other soils. Specific suppression appears to be associated with soilborne bacteria that are rod-shaped, non spore forming, and heat sensitive (16, 63).

Cook and Rovira (16) hypothesized that soil bacteria, particularly the fluorescent pseudomonads, are primarily responsible for specific antagonism in take-all suppressive soils. They based their hypothesis on the following observations: (1) In pot bioassays, only eight of the 100 isolates of bacteria and actinomycetes tested suppressed take-all at levels equal to or greater than a known suppressive soil. These isolates were all pseudomonads, seven of which were fluorescent. (2) Populations of fluorescent pseudomonads recovered from a suppressive soil were higher than populations recovered from a non suppressive soil. Populations of total bacteria, actinomycetes, and aerobic spore-forming bacteria recovered from both soils, however, were similar. (3) In suppressive soils, massive bacterial development occurred in the vicinity of root lesions caused by *G. graminis*. Fungal hyphae were eventually colonized by the bacteria and lysed.

Although several studies have been published on the role of bacteria in the biological control of take-all of wheat (88, 89, 90, 91, 92, 93, 99), little information is available on the bacterial suppression of take-all patch or other turfgrass patch diseases. In 1984, Wong and Baker (99) reported that the development of take-all patch (reported as Ophiobolus patch) in pots was inhibited by fluorescent pseudomonads recovered from a soil suppressive to Fusarium wilt. In treatments consisting of both the antagonist and *G. graminis* var. *avenae*, the severity of take-all patch was less than 10% after 4 wk. Furthermore, symptoms completely disappeared within 6 wk. The authors suggested that fluorescent pseudomonads possess the potential for biological control of take-all patch in golf course turf. Since fluorescent pseudomonads rapidly colonize soils fumigated with methyl bromide (62), a practice often used in golf course greens establishment, they may dominate the soil microflora long enough to inhibit disease development (99). This theory, however, has yet to be confirmed.

Wong and Siviour (100) suggested that the mechanisms responsible for take-all decline in cereals could account for the disappearance of take-all patch in severely diseased bentgrass turf (Plate 58). Landschoot and Clarke (*unpublished data*) observed such a decline in take-all severity over a four year period on a newly constructed golf course in New Jersey.

Cross Protection

Wong (97) defined cross protection as "the protection of a plant from infection by a microbial pathogen following simultaneous or prior exposure to another microorganism." This concept was first established when researchers were able to protect tobacco from a virulent strain of tobacco mosaic virus by first inoculating plants with a less virulent strain of the virus (54). The principle of cross protection has been used in the prevention of several bacterial and fungal diseases (1, 2, 20, 66, 83, 95).

Balis (2), Scott (66), and Deacon (20) were among the first to report that prior colonization of wheat roots with an avirulent *Phialophora* spp. reduced the incidence of take-all. Wong (95) demonstrated that wheat roots pre-colonized by an avirulent strain of *G. graminis* var. *graminis* were protected from subsequent infection by the pathogens *G. graminis* var. *tritici* and *G. graminis* var. *avenae*. In other studies, a reduction in the severity of take-all was achieved using hypovirulent strains of *G. graminis* var. *tritici* (1, 83). Wong (97) stated that avirulent *Gaeumannomyces* and *Phialophora* species have successfully protected plants because they occupy the same ecological niche as *G. graminis* var. *tritici* and *G. graminis* var. *avenae*.

To clarify the mechanisms responsible for cross protection, Speakman and Lewis (80) examined the changes in root structure that accompanied infection by *G. graminis* var. *tritici* and two varieties of *Phialophora*. When inoculated separately, the *Phialophora* fungi, but not *G. graminis* var. *tritici*, elicited increased lignification of stelar tissues and increased lignification and suberization in the endodermis. These structural changes presumably prevented the longitudinal spread of *G. graminis* var. *tritici* when inoculated on the same root as the *Phialophora* fungi. The authors concluded that cross protection is mediated by the host but occurs in response to infection by *P. graminicola*.

When assessing factors that affect the occurrence of take-all patch, Deacon (20) noted that the disease often develops in turf following liming or in recently fumigated soils. Since populations of *Phialophora* spp. are typically absent in these situations, Deacon used this indirect evidence to conclude that *Phialophora* spp. may contribute to the suppression of take-all patch in bentgrass turf. In pot studies, Wong and Siviour (100) created soils suppressive to take-all patch by incorporating either *G. graminis* var. *graminis*, an avirulent isolate of *G. graminis* var. *tritici*, or one of two *Phialophora* isolates into sterilized soil. By mixing these suppressive soils with inoculum of the pathogen at the time of seeding, the development of take-all patch was greatly reduced. The authors discussed the potential of using artificially-developed suppressive soils as a top-dressing to control take-all patch on golf course greens.

Although cross protection of take-all patch with non pathogenic isolates of *Phialophora* and *Gaeumannomyces* has been reported, it is unknown whether other turfgrass patch diseases can be suppressed in a similar manner. Because the ERI fungi are slow to colonize turf, the large scale introduction of avirulent *Phialophora* and *Gaeumannomyces* fungi into turf may not be practical. Furthermore, populations of antagonists may not persist when subjected to the intensive use of pesticides on golf and sports turf areas.

LITERATURE CITED

1. Asher, M. J. C. 1978. Interactions between isolates of *Gaeumannomyces graminis* var. *tritici*. Trans. Br. Mycol. Soc. 71:367-373.
2. Balis, C. 1970. A comparative study of *Phialophora radicicola*, an avirulent fungal root parasite of grasses and cereals. Ann. Appl. Biol. 66:59-73.
3. Bean, G. A. 1966. Observations on Fusarium blight of turfgrasses. Plant Dis. Rep. 50:942-945.

4. Bean, G. A. 1969. The role of moisture and crop debris in the development of Fusarium blight of Kentucky bluegrass. Phytopathology 59:479-481.

5. Beard, J. B. 1973. Turfgrass: Science and Culture. Prentice-Hall, Englewood Cliffs, NJ.

6. Brooks, D. H. 1964. Infection of wheat roots by ascospores of *Ophiobolus graminis*. Nature 203:203.

7. Brooks, D. H. 1965. Root infection by ascospores of *Ophiobolus graminis* as a factor in epidemiology of the take-all disease. Trans. Br. Mycol. Soc. 48:237-248.

8. Brown, M. E., and Hornby, D. 1971. Behaviour of *Ophiobolus graminis* on slides buried in soil in the presence or absence of wheat seedlings. Trans. Br. Mycol. Soc. 56:95-103.

9. Butler, F. C. 1953. Saprophytic behaviour of some cereal root-rot fungi. I. Saprophytic colonization of wheat straw. Ann. Appl. Biol. 40:284-297.

10. Butler, F. C. 1953. Saprophytic behaviour of some cereal root-rot fungi. II. Factors influencing saprophytic colonization of wheat straw. Ann Appl. Biol. 40:298-304.

11. Chastagner, G. A., and Hammer, B. 1987. Current research on necrotic ring spot. Pages 94-95 in: Proc. Northwest Turfgrass Conf., 41st. Northwest Turfgrass Association, Gleneden Beach, OR.

12. Cook, R. J. 1981. The effect of soil reaction and physical conditions. Pages 343-352 in: Biology and Control of Take-all. M. J. C. Asher and P. J. Shipton, eds. Academic Press, New York.

13. Cook, R. J., and Christen, A. A. 1976. Growth of cereal root-rot fungi as affected by temperature-water potential interactions. Phytopathology 66:193-197.

14. Cook, R. J., and Papendick, R. I. 1972. Influence of water potential of soils and plants on root disease. Annu. Rev. Phytopathol. 10:349-374.

15. Cook, R. J., Papendick, R. I., and Griffin, D. M. 1972. Growth of two root-rot fungi as affected by osmotic and

matric water potentials. Soil Sci. Soc. Am. Proc. 36:78-82.

16. Cook, R. J., and Rovira, A. D. 1976. The role of bacteria in the biological control of *Gaeumannomyces graminis* by suppressive soils. Soil Biol. Biochem. 8:269-273.

17. Couch, H. B., and Bedford, E. R. 1966. Fusarium blight of turfgrasses. Phytopathology 56:781-786.

18. Crahay, J. N., Dernoeden, P. H., and O'Neill, N. R. 1988. Growth and pathogenicity of *Leptosphaeria korrae* in bermudagrass. Plant Dis. 72:945-949.

19. Davis, D. B., and Dernoeden, P. H. 1991. Summer patch and Kentucky bluegrass quality as influenced by cultural practices. Agron. J. 83:670-677.

20. Deacon, J. W. 1973. Factors affecting occurrence of the Ophiobolus patch disease of turf and its control by *Phialophora radicicola*. Plant Pathol. 22:149-155.

21. Deacon, J. W. 1976. Biological control of the take-all fungus, *Gaeumannomyces graminis*, by *Phialophora radicicola* and similar fungi. Soil Biol. Biochem. 8:275-283.

22. Dernoeden, P. H., and O'Neill, N. R. 1983. Occurrence of Gaeumannomyces patch disease in Maryland and growth and pathogenicity of the causal agent. Plant Dis. 67:528-532.

23. Elliott, M. L. 1991. Determination of an etiological agent of bermudagrass decline. Phytopathology 81:1380-1384.

24. Endo, R. M. 1972. The turfgrass community as an environment for the development of facultative fungal parasites. Pages 171-202 in: The Biology and Utilization of Grasses. V. B. Youngner and C. M. McKell, eds. Academic Press, New York.

25. Endo, R. M., and Colbaugh, P. F. 1974. Fusarium blight of Kentucky bluegrass in California. Pages 325-327 in: Proc. Int. Turfgrass Res. Conf., 2nd. E. C. Roberts, ed.

American Society of Agronomy and Crop Science Society of America, Madison, WI.

26. Endo, R. M., Ohr, H. D., and Krausman, E. M. 1985. *Leptosphaeria korrae*, a cause of the spring dead spot disease of bermudagrass in California. Plant Dis. 69:235-237.

27. Fellows, H. 1941. Effect of certain environmental conditions on the prevalence of *Ophiobolus graminis* in the soil. J. Agric. Res. 63:715-726.

28. Gams, W., and Domsch, K. H. 1967. Beiträge zur anwendung der Bodenwaschtechnik für die isolierung von Bodenpilzen. Arch. Mikrobiol. 58:134-144.

29. Garrett, S. D. 1937. Soil conditions and the take-all disease of wheat. II. The relation between soil reaction and soil aeration. Ann. Appl. Biol. 24:747-751.

30. Garrett, S. D. 1938. Soil conditions and the take-all disease of wheat. III. Decomposition of the resting mycelium of *Ophiobolus graminis* in infected wheat stubble buried in the soil. Ann. Appl. Biol. 25:742-766.

31. Garrett, S. D. 1939. Soil conditions and the take-all disease of wheat. IV. Factors limiting infection by ascospores of *Ophiobolus graminis*. Ann. Appl. Biol. 26:47-55.

32. Garrett, S. D. 1940. Soil conditions and the take-all disease of wheat. V. Further experiments on the survival of *Ophiobolus graminis* in infected wheat stubble buried in the soil. Ann. Appl. Biol. 27:199-204.

33. Garrett, S. D. 1956. Biology of Root-Infecting Fungi. Cambridge at the University Press, Cambridge.

34. Garrett, S. D. 1970. Pathogenic Root-Infecting Fungi. Cambridge at the University Press, Cambridge.

35. Glynne, M. D. 1935. Incidence of take-all on wheat and barley on experimental plots at Woburn. Ann. Appl. Biol. 22:225-235.

36. Gould, C. J., Goss, R. L., and Eglitis, M. 1961. Ophiobolus patch disease of turf in western Washington. Plant Dis. Rep. 45:296-297.

37. Henry, A. W. 1932. Influence of soil temperature and soil sterilization on the reaction of wheat seedlings to *Ophiobolus graminis* Sacc. Can. J. Res. 7:198-203.

38. Hornby, D. 1969. Methods of investigating populations of the take-all fungus (*Ophiobolus graminis*) in soil. Ann. Appl. Biol. 64:503-513.

39. Hornby, D. 1979. Take-all decline: A theorist's paradise. Pages 133-156 in: Soil-Borne Plant Pathogens. B. Schippers and W. Gams, eds. Academic Press, New York.

40. Hornby, D. 1981. Inoculum. Pages 271-293 in: Biology and Control of Take-all. M. J. C. Asher and P. J. Shipton, eds. Academic Press, New York.

41. Jackson, N. 1980. Gaeumannomyces (Ophiobolus) patch disease. Pages 141-143 in: Advances in Turfgrass Pathology. P. O. Larsen and B. G. Joyner, eds. Harcourt Brace Jovanovich, Duluth, MN.

42. Jackson, N. 1987. Some patch diseases of turfgrasses associated with root and crown infecting fungi. N. Z. J. Turf Management 1:21-23.

43. Kackley, K. E., Grybauskas, A. P., and Dernoeden, P. H. 1990. Growth of *Magnaporthe poae* and *Gaeumannomyces incrustans* as affected by temperature-osmotic potential interactions. Phytopathology 80:646-650.

44. Kackley, K. E., Grybauskas, A. P., Dernoeden, P. H., and Hill, R. L. 1990. Role of drought stress in the development of summer patch in field-inoculated Kentucky bluegrass. Phytopathology 80:655-658.

45. Kackley, K. E., Grybauskas, A. P., Hill, R. L., and Dernoeden, P. H. 1990. Influence of temperature-soil water status interactions on the development of summer patch in *Poa* spp. Phytopathology 80:650-655.

46. Landschoot, P. J., and Jackson, N. 1990. Pathogenicity of some ectotrophic fungi with *Phialophora* anamorphs that infect the roots of turfgrasses. Phytopathology 80:520-526.

47. Lucas, L. T. 1980. Control of spring dead spot of bermudagrass with fungicides in North Carolina. Plant Dis. 64:868-870.

48. Lucas, L. T. 1980. Spring deadspot of bermudagrass. Pages 183-187 in: Advances in Turfgrass Pathology. P. O. Larsen and B. G. Joyner, eds. Harcourt Brace Jovanovich, Duluth, MN.

49. Lucas, R. L. 1955. A comparative study of *Ophiobolus graminis* and *Fusarium culmorum* in saprophytic colonization of wheat straw. Ann. Appl. Biol. 43:134-143.

50. MacNish, G. C. 1973. Survival of *Gaeumannomyces graminis* var. *tritici* in field soil stored in controlled environments. Aust. J. Biol. Sci. 26:1319-1325.

51. McCarty, L. B., DiPaola, J. M., and Lucas, L. T. 1991. Regrowth of bermudagrass infected with spring dead spot following low temperature exposure. Crop Sci. 31:182-184.

52. McCarty, L. B., and Lucas, L. T. 1988. Identification and suppression of spring dead spot disease in bermudagrass. Page 154 in: Agronomy Abstracts. American Society of Agronomy, Madison, WI.

53. McCarty, L. B., and Lucas, L. T. 1989. *Gaeumannomyces graminis* associated with spring dead spot of bermudagrass in the southeastern United States. Plant Dis. 73:659-661.

54. McKinney, H. H. 1929. Mosaic diseases in the Canary Islands, West Africa, and Gibraltar. J. Agric. Res. 39:557-578.

55. McKinney, H. H., and Davis, R. J. 1925. Influence of soil temperature and moisture on infection of young wheat plants by *Ophiobolus graminis*. J. Agric. Res. 31:827-840.

56. Nilsson, H. E., and Smith, J. D. 1981. Take-all of grasses. Pages 433-448 in: Biology and Control of Take-all. M. J. C. Asher and P. J. Shipton, eds. Academic Press, New York.

57. Pair, J. C., Crowe, F. J., and Willis, W. G. 1986. Transmission of spring dead spot disease of bermudagrass by turf/soil cores. Plant Dis. 70:877-878.

58. Papendick, R. I., and Cook, R. J. 1974. Plant water stress and development of Fusarium foot rot in wheat subjected to different cultural practices. Phytopathology 64:358-363.

59. Pope, A. M. S., and Jackson, R. M. 1973. Effects of wheatfield soil on inocula of *Gaeumannomyces graminis* (Sacc.) Arx & Olivier var. *tritici* J. Walker in relation to take-all decline. Soil Biol. Biochem. 5:881-890.

60. Rao, A. S. 1959. A comparative study of competitive saprophytic ability in twelve root-infecting fungi by an agar plate method. Trans. Br. Mycol. Soc. 42:97-111.

61. Reis, E. M., Cook, R. J., and McNeal, B. L. 1981. Effect of plant nutrients on take-all of wheat. (Abstr.) Phytopathology 71:108.

62. Ridge, E. H. 1976. Studies in soil fumigation--II. Effects on bacteria. Soil Biol. Biochem. 8:249-253.

63. Rovira, A. D., and Wildermuth, G. B. 1981. The nature and mechanisms of suppression. Pages 385-415 in: Biology and Control of Take-all. M. J. C. Asher and P. J. Shipton, eds. Academic Press, New York.

64. Samuel, G., and Garrett, S. D. 1933. Ascospore discharge in *Ophiobolus graminis*, and its probable relation to the development of whiteheads in wheat. Phytopathology 23:721-728.

65. Sanford, G. B., and Broadfoot, W. C. 1931. Studies of the effects of other soil-inhabiting micro-organisms on the virulence of *Ophiobolus graminis* Sacc. Sci. Agric. 11:512-528.

66. Scott, P. R. 1970. *Phialophora radicicola*, an avirulent parasite of wheat and grass roots. Trans. Br. Mycol. Soc. 55:163-167.

67. Shipton, P. J. 1981. Saprophytic survival between susceptible crops. Pages 295-316 in: Biology and

Control of Take-all. M. J. C. Asher and P. J. Shipton, eds. Academic Press, New York.

68. Sivasithamparam, K., and Parker, C. A. 1981. Physiology and nutrition in culture. Pages 125-150 in: Biology and Control of Take-all. M. J. C. Asher and P. J. Shipton, eds. Academic Press, New York.

69. Skou, J. P. 1981. Morphology and cytology of the infection process. Pages 175-197 in: Biology and Control of Take-all. M. J. C. Asher and P. J. Shipton, eds. Academic Press, New York.

70. Smiley, R. W. 1980. Fusarium blight of Kentucky bluegrass: New perspectives. Pages 155-175 in: Advances in Turfgrass Pathology. P. O. Larsen and B. G. Joyner, eds. Harcourt Brace Jovanovich, Duluth, MN.

71. Smiley R. W., and Cook, R. J. 1973. Relationship between take-all of wheat and rhizosphere pH in soils fertilized with ammonium vs. nitrate-nitrogen. Phytopathology 63:882-890.

72. Smiley, R. W., and Craven Fowler, M. 1984. *Leptosphaeria korrae* and *Phialophora graminicola* associated with Fusarium blight syndrome of *Poa pratensis* in New York. Plant Dis. 68:440-442.

73. Smiley, R. W., Craven Fowler, M., and O'Knefski, R. C. 1985. Arsenate herbicide stress and incidence of summer patch on Kentucky bluegrass turfs. Plant Dis. 69:44-48.

74. Smiley, R. W., Dernoeden, P. H., and Clarke, B. B. 1992. Compendium of Turfgrass Diseases. 2nd ed. American Phytopathological Society, St. Paul, MN.

75. Smiley, R. W., Fowler, M. C., and Kane, R. T. 1985. Temperature and osmotic potential effects on *Phialophora graminicola* and other fungi associated with patch diseases of *Poa pratensis*. Phytopathology 75:1160-1167.

76. Smith, A. M. 1965. *Ophiobolus herpotrichus*, a cause of spring dead spot in couch turf. Agric. Gaz. N.S.W. 76:753-758.

77. Smith, A. M. 1971. Control of spring dead spot of couch grass turf in New South Wales. J. Sports Turf Res. Inst. 47:60-65.

78. Smith, J. D. 1956. Fungi and Turf Diseases. (6) Ophiobolus patch disease. J. Sports Turf Res. Inst. 9:180-202.

79. Smith, J. D., Jackson, N., and Woolhouse, A. R. 1989. Fungal Diseases of Amenity Turf Grasses. E. & F. N. Spon, London.

80. Speakman, J. B., and Lewis, B. G. 1978. Limitation of *Gaeumannomyces graminis* by wheat root responses to *Phialophora radicicola*. New Phytol. 80:373-380.

81. Stuckey, I. H. 1941. Seasonal growth of grass roots. Am. J. Bot. 28:486-491.

82. Tisserat, N. A., Pair, J. C., and Nus, A. 1989. *Ophiosphaerella herpotricha*, a cause of spring dead spot of bermudagrass in Kansas. Plant Dis. 73:933-937.

83. Tivoli, B., Férault, A. C., Lemaire, J. M., and Spire, D. 1979. Agressivité et particules de type viral dans huit isolats monoascosporés de *Gaeumannomyces graminis* (Sacc.) Arx et Olivier (*Ophiobolus graminis* Sacc.). Ann. Phytopathol. 11:259-261.

84. Walker, J. 1975. Take-all diseases of Gramineae: A review of recent work. Rev. Plant Pathol. 54:113-144.

85. Walker, J. 1981. Taxonomy of take-all fungi and related genera and species. Pages 15-74 in: Biology and Control of Take-all. M. J. C. Asher and P. J. Shipton, eds. Academic Press, New York.

86. Walker, J., and Smith, A. M. 1972. *Leptosphaeria narmari* and *L. korrae* spp. nov., two long-spored pathogens of grasses in Australia. Trans. Br. Mycol. Soc. 58:459-466.

87. Warcup, J. H. 1957. Studies on the occurrence and activity of fungi in a wheat-field soil. Trans. Br. Mycol. Soc. 40:237-262.

88. Weller, D. M. 1983. Colonization of wheat roots by a fluorescent pseudomonad suppressive to take-all. Phytopathology 73:1548-1553.

89. Weller, D. M. 1985. Application of fluorescent pseudomonads to control root diseases. Pages 137-140 in: Ecology and Management of Soilborne Plant Pathogens. C. A. Parker, A. D. Rovira, K. J. Moore, P. T. W. Wong, and J. F. Kollmorgan, eds. American Phytopathological Society, St. Paul, MN.

90. Weller, D. M. 1988. Biological control of soilborne plant pathogens in the rhizosphere with bacteria. Annu. Rev. Phytopathol. 26:379-407.

91. Weller, D. M., and Cook, R. J. 1983. Suppression of take-all of wheat by seed treatments with fluorescent pseudomonads. Phytopathology 73:463-469.

92. Weller, D. M., Howie, W. J., and Cook, R. J. 1988. Relationship between in vitro inhibition of *Gaeumanno-myces graminis* var. *tritici* and suppression of take-all of wheat by fluorescent pseudomonads. Phytopathology 78:1094-1100.

93. Weller, D. M., Zhang, B.-X., and Cook, R. J. 1985. Application of a rapid screening test for selection of bacteria suppressive to take-all of wheat. Plant Dis. 69:710-713.

94. Weste, G. 1972. The process of root infection by *Ophiobolus graminis*. Trans. Br. Mycol. Soc. 59:133-147.

95. Wong, P. T. W. 1975. Cross-protection against the wheat and oat take-all fungi by *Gaeumannomyces graminis* var. *graminis*. Soil Biol. Biochem. 7:189-194.

96. Wong, P. T. W. 1980. Effect of temperature on growth of some avirulent fungi and cross-protection against the wheat take-all fungus. Ann. Appl. Biol. 95:291-299.

97. Wong, P. T. W. 1981. Biological control by cross-protection. Pages 417-431 in: Biology and Control of Take-all. M. J. C. Asher and P. J. Shipton, eds. Academic Press, New York.

98. Wong, P. T. W. 1983. Effect of osmotic potential on the growth of *Gaeumannomyces graminis* and *Phialophora* spp. Ann. Appl. Biol. 102:67-78.

99. Wong, P. T. W., and Baker, R. 1984. Suppression of wheat take-all and Ophiobolus patch by fluorescent pseudomonads from a Fusarium-suppressive soil. Soil Biol. Biochem. 16:397-403.

100. Wong, P. T. W., and Siviour, T. R. 1979. Control of Ophiobolus patch in *Agrostis* turf using avirulent fungi and take-all suppressive soils in pot experiments. Ann. Appl. Biol. 92:191-197.

101. Worf, G. L., Stewart, J. S., and Avenius, R. C. 1986. Necrotic ring spot disease of turfgrass in Wisconsin. Plant Dis. 70:453-458.

102. Youngner, V. B. 1969. Physiology of growth and development. Pages 187-216 in: Turfgrass Science. A. A. Hanson and F. V. Juska, eds. American Society of Agronomy, Madison, WI.

1. Spring dead spot, caused by *Leptosphaeria korrae,* on a *Cynodon dactylon* lawn. (Courtesy N. Jackson)

2. Severe damage resulting from spring dead spot on a *Cynodon dactylon* lawn. (Courtesy N. Jackson)

3. Spring dead spot of *Cynodon dactylon*. (Courtesy P. H. Dernoeden)

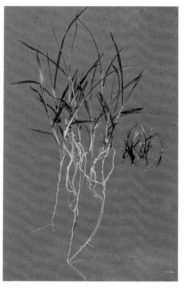

4. Growth of *Cynodon dactylon* seedlings in the presence (right) or absence (left) of *Leptosphaeria korrae*. (Reprinted from Compendium of Turfgrass Diseases, 2nd ed., APS Press, 1992*)

5. Craterlike depressions in *Poa pratensis* turf caused by *Leptosphaeria korrae*. (Courtesy N. Jackson)

* Compendium of Turfgrass Diseases, 2nd ed., by R. W. Smiley, P. H. Dernoeden, and B. B. Clarke, American Phytopathological Society, St. Paul, MN, 1992.

6. Symptoms of necrotic ring spot, caused by *Lepto-sphaeria korrae*, on *Poa pratensis*. (Courtesy N. Jackson)

7. Necrotic ring spot symptoms on a 2-year-old *Poa pratensis* lawn. (Courtesy N. Jackson)

8. Necrotic ring spot on a *Poa pratensis* lawn. (Courtesy B. B. Clarke)

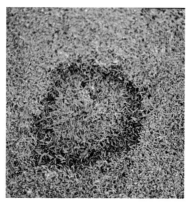

9. Regrowth of *Agrostis palustris* within the center of a necrotic ring spot on a *Poa annua* green. (Courtesy B. B. Clarke)

10. Necrotic ring spot on a *Festuca rubra* fairway. (Courtesy B. B. Clarke)

11. Necrosis of roots and rhizomes of *Poa pratensis* caused by *Leptosphaeria korrae*. (Reprinted from Compendium of Turfgrass Diseases, 2nd ed., APS Press, 1992)

12. Crown rot of *Poa pratensis* infected by *Leptosphaeria korrae*. (Courtesy W. W. Shane)

13. Pseudothecia of *Leptosphaeria korrae* embedded in a leaf sheath of *Poa pratensis*. (Reprinted from Compendium of Turfgrass Diseases, 2nd ed., APS Press, 1992)

14. Exposed pseudothecia of *Leptosphaeria korrae* on wheat (*Triticum aestivum*). (Courtesy G. L. Worf)

15. Pseudothecium of *Leptosphaeria korrae* on a decomposing crown of *Poa pratensis*. (Reprinted from Compendium of Turfgrass Diseases, 2nd ed., APS Press, 1992)

16. Ascospores of *Leptosphaeria korrae*. (Reprinted from Compendium of Turfgrass Diseases, 2nd ed., APS Press, 1992)

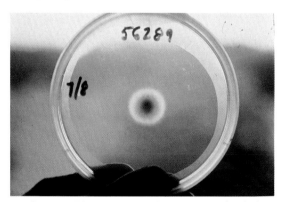

17. Culture of *Leptosphaeria korrae* grown for 3 days on potato-dextrose agar at 21 C. (Courtesy W. W. Shane)

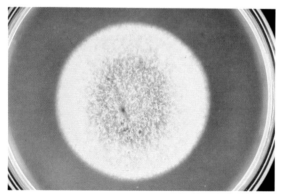

18. Culture of *Leptosphaeria korrae* grown for 10 days on potato-dextrose agar at 21 C. (Courtesy P. J. Landschoot)

19. Perithecia of *Gaeumannomyces incrustans* on wheat (*Triticum aestivum*). (Courtesy P. J. Landschoot)

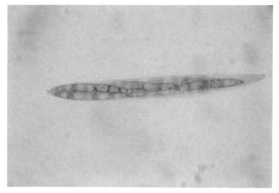

20. Ascus and ascospores of *Gaeumannomyces incrustans* stained with 1% phloxine. (Courtesy P. J. Landschoot)

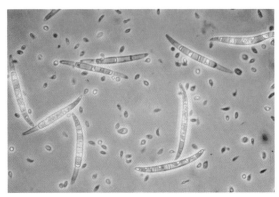

21. Ascospores and phialospores (smaller spores) of *Gaeumannomyces incrustans*. (Courtesy P. J. Landschoot)

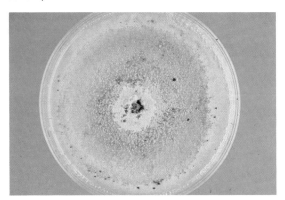

22. Culture of *Gaeumannomyces incrustans* grown for 7 days on potato-dextrose agar at 28 C. (Courtesy P. J. Landschoot)

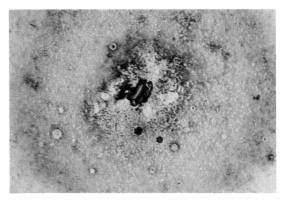

23. Culture of *Gaeumannomyces incrustans* exhibiting characteristic black mycelial crusts on potato-dextrose agar after 25 days at 28 C. (Courtesy P. J. Landschoot)

24. Devastation of the South Lawn of the White House by Fusarium blight evident during an address by President Lyndon Johnson, August 1964. (Courtesy LBJ Library, Austin, TX)

25. Early symptoms of summer patch on *Poa pratensis* caused by *Magnaporthe poae*. (Courtesy R. W. Smiley)

26. Summer patch on *Poa pratensis*. Note regrowth within the center of the patch. (Reprinted from Compendium of Turfgrass Diseases, 2nd ed., APS Press, 1992)

27. Growth of *Lolium perenne* within a patch of *Poa pratensis* killed by *Magnaporthe poae*. (Courtesy B. B. Clarke)

28. Typical frogeye symptoms of summer patch on a *Poa annua* fairway. (Courtesy P. J. Landschoot)

29. Wilting of *Poa pratensis*, infected by *Magnaporthe poae*, during heat stress. (Courtesy N. Jackson)

30. Symptoms of summer patch coalescing on *Poa pratensis* sod. (Courtesy N. Jackson)

31. Severe damage resulting from summer patch on a *Poa pratensis* lawn. (Courtesy R. W. Smiley)

32. Summer patch on a *Poa pratensis* golf fairway. (Reprinted from Compendium of Turfgrass Diseases, 2nd ed., APS Press, 1992)

33. Leaf lesions associated with summer patch of *Poa pratensis*. (Courtesy N. Jackson)

34. Summer patch of *Festuca rubra* caused by *Magnaporthe poae*. (Reprinted from Compendium of Turfgrass Diseases, 2nd ed., APS Press, 1992)

35. Summer patch symptoms on a golf green. Note *Agrostis palustris* surviving within patches of dying *Poa annua*. (Courtesy P. H. Dernoeden)

36. Symptoms of summer patch accentuated by soil compaction on a *Poa annua* golf green. (Courtesy P. J. Landschoot)

37. Root and crown discoloration of *Poa pratensis* caused by *Magnaporthe poae*. (Courtesy P. J. Landschoot)

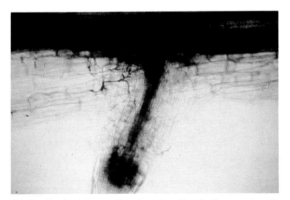

38. Cortical and vascular discoloration in *Poa pratensis* roots colonized by *Magnaporthe poae*. (Courtesy P. J. Landschoot)

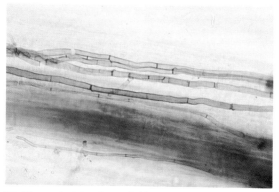

39. Pigmented hyphae of *Magnaporthe poae* on the surface of a root. (Reprinted from Compendium of Turf-grass Diseases, 2nd ed., APS Press, 1992)

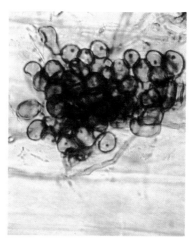

40. Slightly lobed hyphopodia of *Magnaporthe poae*. (Courtesy P. J. Landschoot)

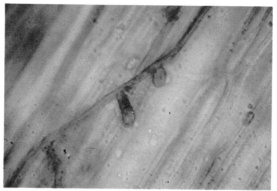

41. Simple hyphopodia of *Phialophora graminicola*. (Courtesy N. Jackson)

42. Growth cessation structures of *Phialophora graminicola.* (Courtesy P. J. Landschoot)

43. Perithecia of *Magnaporthe poae* on a wheat (*Triticum aestivum*) stem. (Reprinted from Compendium of Turfgrass Diseases, 2nd ed., APS Press, 1992)

44. Ascus and ascospores of *Magnaporthe poae.* Note refractive ring at apex of ascus. (Courtesy P. J. Landschoot)

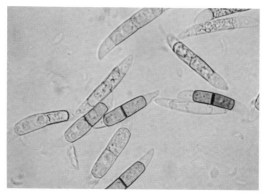

45. Ascospores and phialospores (smaller spores) of *Magnaporthe poae*. (Reprinted from Compendium of Turfgrass Diseases, 2nd ed., APS Press, 1992)

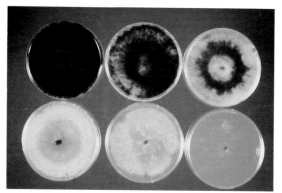

46. Cultures of *Magnaporthe poae* grown for 26 days on (counterclockwise from top right) 0.25, 0.5, 1.0, 0.12, and 0.06 strength potato-dextrose agar. Culture on the lower right is *M. poae* grown on water agar for the same period. (Courtesy R. W. Smiley)

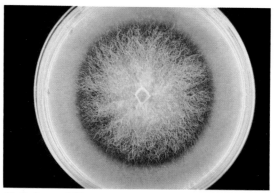

47. Culture of *Magnaporthe poae* grown for 6 days on half-strength potato-dextrose agar at 28 C. (Courtesy P. J. Landschoot)

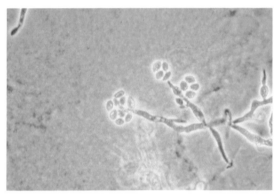

48. Phialospores of the anamorphic state (*Phialophora*) of *Magnaporthe poae*. (Reprinted from Compendium of Turfgrass Diseases, 2nd ed., APS Press, 1992)

49. Early symptoms of take-all patch on a 3-week-old *Agrostis stolonifera* green. (Courtesy B. B. Clarke)

50. Symptoms of take-all patch, caused by *Gaeumannomyces graminis* var. *avenae*, on an *Agrostis stolonifera* green. (Courtesy B. B. Clarke)

51. Take-all patch on an *Agrostis* fairway. (Courtesy P. J. Landschoot)

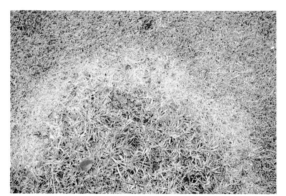

52. Take-all patch symptoms on *Agrostis tenuis* caused by *Gaeumannomyces graminis* var. *avenae*. (Courtesy N. Jackson)

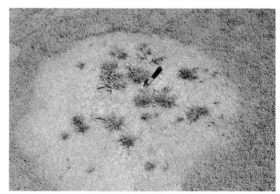

53. Take-all patch on an *Agrostis fairway* with the center of the patch recolonized by *Festuca rubra*. (Courtesy B. B. Clarke)

54. Take-all patch of an *Agrostis* golf fairway. Note centers of patches have been recolonized by *Poa annua*. (Courtesy R. W. Smiley)

55. Take-all patch of an *Agrostis* golf fairway. Note recolonization of patch centers by *Poa pratensis*. (Reprinted from Compendium of Turfgrass Diseases, 2nd ed., APS Press, 1992)

56. Take-all patch of an *Agrostis palustris* golf fairway. (Courtesy P. H. Dernoeden)

57. Symptoms of take-all patch accentuated by environmental stress on an *Agrostis* golf green. (Courtesy P. H. Dernoeden)

58. Take-all decline on an *Agrostis* golf green. Note faint symptoms from previous infections. (Courtesy P. H. Dernoeden)

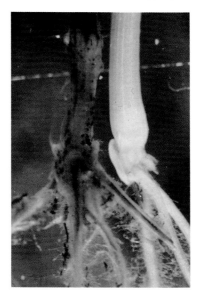

59. Comparison of a necrotic (left) and healthy (right) wheat (*Triticum aestivum*) seedling infected with *Gaeumannomyces graminis* var. *avenae*. (Courtesy N. Jackson)

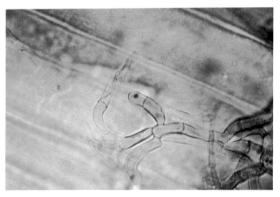

60. Simple hyphopodia of *Gaeumannomyces graminis* var. *avenae*. (Reprinted from Compendium of Turfgrass Diseases, 2nd ed., APS Press, 1992)

61. Perithecia of *Gaeumannomyces graminis* var. *avenae* on an *Agrostis* tiller. (Reprinted from Compendium of Turfgrass Diseases, 2nd ed., APS Press, 1992)

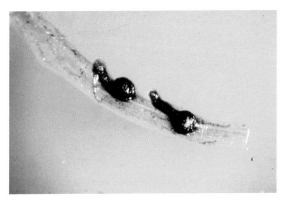

62. Perithecia of *Gaeumannomyces graminis* var. *avenae* on wheat (*Triticum aestivum*). (Courtesy P. J. Landschoot)

63. Perithecium of *Gaeumannomyces graminis* var. *avenae* extruding asci. (Courtesy P. H. Dernoeden)

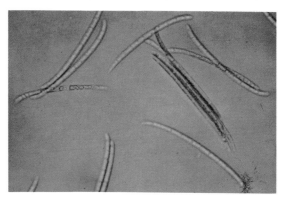

64. Ascospores of *Gaeumannomyces graminis* var. *avenae*. (Courtesy N. Jackson)

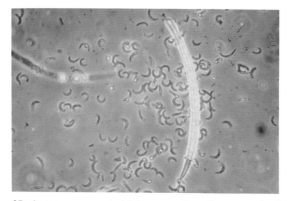

65. Ascospores and phialospores (smaller spores) of *Gaeumannomyces graminis* var. *avenae*. (Courtesy P. J. Landschoot)

66. Bermudagrass decline on a *Cynodon* green. Note the absence of disease on collar that is mowed higher than the putting green. (Reprinted from Compendium of Turfgrass Diseases, 2nd ed., APS Press, 1992)

67. Thinning associated with bermudagrass decline of *Cynodon* spp. turf. (Reprinted from Compendium of Turfgrass Diseases, 2nd ed., APS Press, 1992)

68. Healthy roots of *Cynodon* spp. and roots affected by bermudagrass decline. (Reprinted from Compendium of Turfgrass Diseases, 2nd ed., APS Press, 1992)

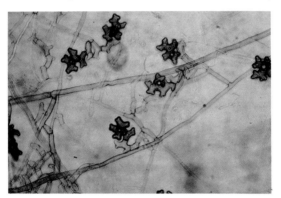

69. Lobed hyphopodia of *Gaeumannomyces graminis* var. *graminis*. (Reprinted from Compendium of Turfgrass Diseases, 2nd ed., APS Press, 1992)

70. Lobed hyphopodia of *Gaeumannomyces graminis* var. *graminis*. (Courtesy P. J. Landschoot)

CURRENT CONCEPTS IN DISEASE DETECTION

William W. Shane
Michigan State University
Southwest Michigan Research & Extension Center
Benton Harbor, MI 49022

and

John C. Stier and **Stephen T. Nameth**
The Ohio State University
Columbus, OH 43210

INTRODUCTION

In recent years, at least seven root-inhabiting pathogens have been associated with patch diseases on amenity turfgrasses (9, 14, 17, 18, 21, 30, 36, 38, 39). These fungi include *Gaeumannomyces graminis* (Sacc.) Arx. & D. Olivier var. *avenae* (E. M. Turner) Dennis; *G. graminis* (Sacc.) Arx. & D. Olivier var. *graminis*; *G. incrustans* Landschoot & Jackson; *Leptosphaeria korrae* J. C. Walker & A. M. Sm.; *L. narmari* J. C. Walker & A. M. Sm.; *Magnaporthe poae* Landschoot & Jackson; and *Ophiosphaerella herpotricha* (Fr.:Fr.) J. C. Walker. Investigations are currently underway to determine the exact distribution of these organisms throughout the United States and to evaluate the factors that influence disease development. As this work progresses, a clearer understanding of the strategies necessary to identify and manage patch diseases is starting to emerge.

In this chapter, the approaches used to detect and identify ectotrophic root-infecting (ERI) are addressed. In particular, the symptoms and signs required for an accurate and rapid identification of patch diseases have been highlighted. The utility of several relatively new identification techniques (e.g., immunoassays, DNA probes, and protein gel electrophoresis) is also presented.

TRADITIONAL TECHNIQUES USED IN DISEASE IDENTIFICATION

Symptom Expression

Since several of the ERI fungi infect different turfgrass species, the identification of the host plant should be the first step in the diagnostic process. The time of year during which the disease occurs and the prevailing environmental conditions are also useful diagnostic features. Although the general appearance of the disease in the field may be helpful in distinguishing patch diseases from other disorders, most patch diseases exhibit similar visual symptoms (e.g., frogeye or donut-shaped patches). In addition, environmental conditions may alter symptom expression. Management practices, differences in cultivar susceptibility, and the age of the sward may also affect the appearance and prevalence of these diseases (31, 33). For a thorough discussion of symptom expression and disease development in the field, refer to Jackson (Chapter two).

Fungal Signs

The ERI fungi possess morphological characteristics that are useful in identification. Microscopic examination of these features (e.g., fruiting bodies, spores, and hyphae) on infected roots and crowns can be enhanced by utilizing clearing (with potassium hydroxide) and staining procedures such as the one described in Fig. 5.1. All members of the ERI fungi that infect turf produce dark brown, ectotrophic runner hyphae, and most produce hyphopodia (32, 37). The presence of runner hyphae, hyphopodia, and growth cessation structures can be utilized to help distinguish the ERI fungi from other darkly-pigmented species that are often observed on turfgrass stems, crowns, and roots. Some of the more common species within this group include *Alternaria, Bipolaris, Cladosporium, Curvularia, Drechslera, Epicoccum,* and *Stemphyllium* (31).

The shape, frequency, and distribution of hyphopodia on host tissue is another useful taxonomic criterion (see Landschoot, Chapter three). In particular, the presence or

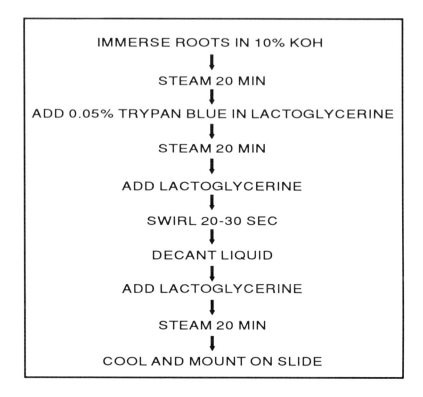

Fig. 5.1. Root clearing method for observing fungal structures on turfgrass roots (modified from Phillips and Hayman (25)).

absence of hyphopodial lobes is often used to differentiate between species of the ERI fungi. For example, *M. poae* produces slightly lobed hyphopodia (Plate 40) on an infrequent basis (18), whereas *G. graminis* var. *graminis* produces abundant, strongly lobed hyphopodia (37) (Plates 69 and 70). Care must also be taken not to rely on any one morphological characteristic, since many of the ERI fungi and dematiacious saprophytes may have similar taxonomic features. The initial infection caused by *G. graminis* var. *avenae*, for instance, often results in darkened crown, rhizome, stolon, root, and stele tissues (Plate 59). This characteristic could be confused with the necrosis induced by the anthracnose pathogen *Colletotrichum graminicola* if that were the only feature considered.

Isolation and Characterization in Culture

Since the ERI fungi are relatively slow-growing in culture, the use of small (3-5 mm) pieces of infected host tissue in the isolation process helps to minimize the occurrence of fast-growing contaminants. The use of semi-selective media (15) and trap hosts (30) may also be used to isolate these pathogens from severely contaminated tissue.

Several procedures have been specifically developed to isolate ERI fungi from infected turfgrass. The success of these procedures depends on the tolerance of the ERI fungi to disinfectants such as sodium hypochlorite and silver nitrate, and to dehydration. In a procedure developed by Worf et al (39), 5-10 mm sections of infected tissue are rinsed with sterile water for 15-30 min. The sections are then surface-sterilized for 30 s in a 1:1 solution of 1% sodium hypochlorite and 95% ethanol, rinsed in sterile distilled water for 1 h, and blotted dry in filter paper. The plant material is then dried for 18-24 h at room temperature and plated on potato dextrose agar amended with 50 mg^{-L} novobiocin.

Landschoot (16) has successfully isolated ERI fungi from host tissue using a modification of the methods developed by

Chambers and Flentje (4) and Davies (6). In this procedure, small (3-5 mm) sections of root or crown tissue are surface-sterilized in a 1% silver nitrate solution for 30-60 s, followed by 5% sodium chloride for 10 s. The tissue is rinsed in sterile, distilled water, blotted dry with clean paper toweling, and then plated on a *Gaeumannomyces*-selective medium (15) or on half-strength potato dextrose agar.

Crahay et al (5) have developed an effective method for the isolation of *L. korrae*. In this method, infected tissue is washed in running tap water for 1 h and then soaked in sterile water for an additional hour. Sections of tissue are then surface-sterilized in a 1:1 solution of 95% alcohol and 1% sodium hypochlorite for 30 s, and soaked for 1 h in sterile water. The roots are dried in sterile filter paper for 24 h and plated on potato dextrose agar amended with 100 mg^{-L} ampicillin.

The color and morphology of the ERI fungi in culture are useful taxonomic features, but they are not usually sufficient to differentiate between species. For instance, both *G. graminis* var. *avenae* and *G. graminis* var. *graminis* produce white to gray, aerial mycelia that curl back toward the colony center on potato dextrose agar (Table 3.3a). Positive identification of the ERI fungi, therefore, often requires examination of the sexual stage.

Several species, including *G. graminis* var. *graminis* and *O. herpotricha*, will produce perithecia on host tissue in the field. Other species such as *M. poae* will produce the sexual stage only in culture (Plate 43). This process, however, is time consuming and is not suitable for rapid diagnostic purposes. In addition, *G. incrustans* and *M. poae*, two heterothallic species, require the presence of the opposite mating type for the production of perithecia. Sexual characteristics of several ERI fungi are described in Tables 3.2a and 3.2b.

APPLICATION OF BIOTECHNOLOGY TO PATCH DISEASE IDENTIFICATION

Although the teleomorph of most ERI fungi can be produced in culture after an extended period of time (6-8 wk), other techniques are now available to aid in identification. Recent advances in biotechnology have resulted in the development of several methodologies that may have direct application for detection and identification of the ERI fungi. They include, but are not limited to, immunoassays, DNA probes, and protein gel electrophoresis.

Immunoassays

Immunoassays have been used for several years to successfully detect fungi in plant tissues (7, 8, 10, 20). Commercial immunoassays were first marketed for the detection of turfgrass pathogens in 1986 with the development of antibodies to *Pythium* spp. (22, 23). Assays developed to detect pathogenic species of *Rhizoctonia*, *Lanzia*, and *Moellerodiscus* were later produced for the turfgrass market (23). These immunoassays have several features in common that make them extremely useful to the turfgrass manager. First, they are rapid and may be used to detect the pathogen in infected swards in approximately 10 min (23). Second, they may be used directly in the field without extensive laboratory facilities or equipment. Third, they have been used successfully to assess and monitor pathogen levels in both asymptomatic and symptomatic turf (23, 28). Unfortunately, such assays are not currently available for the detection of the ERI fungi. In addition, they can be expensive to use and are pathogen specific. For example, an assay specific for *Rhizoctonia* will only detect pathogenic species within that genus. If another incitant is present, the assay will give a negative response.

Immunoassays may utilize both polyclonal and monoclonal antibodies. Polyclonal antibodies are derived from the blood serum of immunized animals and have an affinity for a mixture of antigens or fungal proteins. Monoclonal antibodies are produced by hybridizing the spleen cells of immunized animals with myeloma cells. The resulting "hybridomas" produce antibodies that are specific for a single, target antigen (10, 19). The challenge in both polyclonal and monoclonal antibody production is to develop high concentrations of stable antibodies with the desired specificities.

Although the production of polyclonal antibodies takes a relatively short period (2-3 mo), the supply of antibody is limited by the size and life span of the immunized animal. Once the animal dies, the entire process must be reinitiated. The development of monoclonal antibodies, however, is a labor-intensive process that may take six months to conduct and requires specialized equipment and stringent (sterile) technique. Unlike polyclonal antibodies, monoclonal antibodies can be produced in an essentially unlimited supply.

In the laboratories of the authors, monoclonal antibodies specific for *L. korrae*, the causal agent of necrotic ring spot and an incitant of spring dead spot, have recently been developed (24, 29). Monoclonal antibodies produced from hybridoma clone LKc50 have reacted with lyophilized and non lyophilized mycelia from the more than 50 isolates of *L. korrae* tested. Using an indirect enzyme-linked immunosorbent assay (ELISA), absorbance readings for these isolates were similar to the positive control (*L. korrae* strain ATCC 56289). Readings for two isolates of *O. herpotricha*, however, were similar to the negative controls (selected data presented in Table 5.1). These antibodies have not reacted with turfgrass tissue or other turfgrass pathogens (i.e., *C. graminicola*, *Gaeumannomyces* spp., *M. poae*, *O. herpotricha*, and *Pythium* spp).

Monoclonal antibodies have also been used by the authors to detect *L. korrae* within infected plant tissue (24). Antigen preparations derived from tissue suspected of being infected with *L. korrae* were combined with monoclonal antibodies from clone LKc50 in an indirect ELISA. Absorbance readings for infected samples from Ohio, Idaho, and Washington were more

Table 5.1. Identification of *L. korrae* in culture using monoclonal antibody clone LKc50.

Isolate	Absorbance (405 nm)
L. korrae	
12089B1B	0.620[Y]
12389B	0.539
12389C1	0.754
89-0318A	0.441
O. herpotricha	
27	0.153
67	0.166
Positive control[Z]	0.782
Agar only	0.102
Buffer only	0.000

[Y] Mean of two replicates.
[Z] Homogenized, lyophilized mycelia of *L. korrae* strain ATCC 56289 used at a concentration of 77.4 µg/well.

than three times higher than readings from healthy turf (Fig. 5.2). The intensity of the absorbance reading corresponds to an estimation of the *L. korrae* concentration within the sample (24). This work supports the concept that immunoassays have great potential for the detection and monitoring of ERI fungi within infected turf. Whether this can be done on a commercial basis, however, is unknown.

DNA Probes

The utilization of molecular techniques in disease diagnostics includes the use of probes that are specific for certain DNA sequences within a given organism. These techniques are based on the hybridization of a DNA probe with homologous DNA sequences from the target agent. Depending upon the level of sensitivity required, DNA probes can be utilized in one of several different ways.

Restriction fragment length polymorphism (RFLP) analysis, in which DNA restriction enzyme profiles are combined with specific DNA probes, can reveal differences between species, strains, or isolates of a pathogen. This technique depends upon the fortuitous selection of the proper restriction enzyme(s) and the specific probe. Although RFLP analysis has been used successfully to distinguish between isolates of *O. herpotricha* and *L. korrae* in culture (34), it cannot be used to detect the pathogen within infected plants.

In a second potential application of DNA probes called dot blot hybridization, a DNA sequence is identified that is

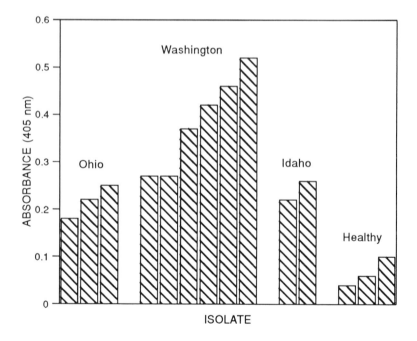

Fig. 5.2. Reaction of *L. korrae*-specific monoclonal antibodies (clone LKc50) to infected Kentucky bluegrass field samples obtained from three states using indirect ELISA.

specific to the target organism. Once this sequence is identified, it can be hybridized with crude DNA preparations of the isolates that are to be tested. In a clinical situation, this technique can be used to detect the presence of target organisms within infected plant tissue.

A third method for detection of turfgrass pathogens is the polymerase chain reaction (PCR). This technique also requires identification of DNA sequences specific for the target organism. Unlike other DNA probe techniques that are based upon hybridization, however, this method is based on the progressive amplification of specific DNA sequences in the target organism. DNA amplification and the detection of positive reactions requires a period of three to four hours. As a pathogen detection technique, PCR has the potential for extreme sensitivity.

DNA probes have recently been used to successfully differentiate isolates of *Gaeumannomyces* spp. (3, 12, 13) and *M. poae* (26, B. I. Hillman, *personal communication*) from other turfgrass pathogens. DNA probes have also been used to detect *L. korrae* in Kentucky bluegrass (*Poa pratensis* L.) and bermudagrass (*Cynodon dactylon* (L.) Pers.) roots (35) and to identify *O. herpotricha* in field samples of bermudagrass showing symptoms of spring dead spot (27). Currently, DNA probes are primarily research tools due to the cost and expertise required to perform these techniques. When further developed and refined, however, DNA probes may be routinely used in a diagnostic setting. These techniques also have the potential for use in epidemiological and ecological studies of turfgrass pathogens.

Protein Gel Electrophoresis

Plant pathologists have utilized protein gel electrophoresis for many years to identify and distinguish between plant pathogens (1, 11). Characteristics such as protein size and charge have been used as the basis for generating protein profiles or "fingerprints" for target organisms. For example, isolates of *G. graminis* have been distinguished on the basis of electrophoretic patterns of proteins extracted from mycelial homogenates (1). Hawkes and Harding (11) utilized isoelectric focusing, a technique that separates proteins based on their electric charge, to identify three species of *Leptosphaeria*.

Two-dimensional electrophoresis and isozyme analysis are two electrophoretic techniques that have potential for the identification of turfgrass fungi. These techniques may be particularly useful for identification of pathogens that do not readily fruit in culture.

Two-dimensional electrophoresis separates proteins by size and charge, resulting in protein profiles that have a high degree of resolution. This technique has been used successfully to distinguish between pathotypes of nematodes (2) and has the potential to distinguish between strains or isolates of the same fungal species. In isozyme analysis, the size of protein enzymes (isozymes) produced by different isolates are compared. Once a series of isozyme patterns are established for a specific organism, isozyme analysis may be used to identify the organism *in vitro*. Both two-dimensional electrophoresis and isozyme analysis are costly, time consuming, and require protein preparations from pure cultures of the target organisms. Although they have potential for use as diagnostic tools, they are not appropriate at this time for rapid detection in a clinical setting or in the field.

SUMMARY

To efficiently detect and identify disease causing agents, the diagnostician must rely on visual symptoms, colony characteristics, microscopic features, and laboratory assays. Immunoassays, DNA probes, and protein gel electrophoresis have the potential to complement standard isolation and detection techniques. The application of these advanced procedures for detection of the ERI fungi, however, is just beginning. Morphological characteristics should continue to represent the primary basis for the classification of the ERI fungi. Continued work with patch diseases and their incitants, however, will undoubtedly result in improved diagnostic methods.

ACKNOWLEDGMENTS

We wish to extend thanks to Dr. Donald Kobayashi and Dr. Bradley Hillman for extensive review of this manuscript.

LITERATURE CITED

1. Abbott, L. K., and Holland, A. A. 1975. Electrophoretic patterns of soluble proteins and isoenzymes of *Gaeumannomyces graminis*. Aust. J. Bot. 23:1-12.
2. Bakker, J., and Bouwman-Smits, L. 1988. Genetic variation in polypeptide maps of two *Globodera rostochiensis* pathotypes. Phytopathology 78:894-900.
3. Bateman, G. L., Ward, E., and Antoniw, J. F. 1992. Identification of *Gaeumannomyces graminis* var. *tritici* and *G. graminis* var. *avenae* using a DNA probe and non-molecular methods. Mycol. Res. 96:737-742.
4. Chambers, S. C., and Flentje, N. T. 1967. Studies on oat-attacking and wheat-attacking isolates of *Ophiobolus graminis* in Australia. Aust. J. Biol. Sci. 20:927-940.
5. Crahay, J. N., Dernoeden, P. H., and O'Neill, N. R. 1988. Growth and pathogenicity of *Leptosphaeria korrae* in bermudagrass. Plant Dis. 72:945-949.
6. Davies, F. R. 1935. Superiority of silver nitrate over mercuric chloride for surface sterilization in the isolation of *Ophiobolus graminis* Sacc. Can. J. Res. 13:168-173.
7. Dewey, F. M., and Brasier, C. M. 1988. Development of ELISA for *Ophiostoma ulmi* using antigen-coated wells. Plant Pathol. 37:28-35.
8. El-Nashaar, H. M., Moore, L. W., and George, R. A. 1986. Enzyme-linked immunosorbent assay quantification of initial infection of wheat by *Gaeumannomyces*

graminis var. *tritici* as moderated by biocontrol agents. Phytopathology 76:1319-1322.

9. Endo, R. M., Ohr, H. D., and Krausman, E. M. 1985. *Leptosphaeria korrae*, a cause of the spring dead spot disease of bermudagrass in California. Plant Dis. 69:235-237.

10. Halk, E. L., and De Boer, S. H. 1985. Monoclonal antibodies in plant-disease research. Annu. Rev. Phytopathol. 23:321-350.

11. Hawkes, N. J., and Harding, H. W. J. 1985. Isoelectric focusing as an aid to the identification of *Leptosphaeria narmari*, a cause of spring dead spot in turf. Australas. Plant Pathol. 14:72-76.

12. Hensen, J. M. 1989. DNA probe for identification of the take-all fungus, *Gaeumannomyces graminis*. Appl. Environ. Microbiol. 55:284-288.

13. Hensen, J. M. 1992. DNA hybridization and polymerase chain reaction (PCR) tests for identification of *Gaeumannomyces*, *Phialophora*, and *Magnaporthe* isolates. Mycol. Res. 96:629-636.

14. Jackson, N. 1981. Take-all patch (Ophiobolus patch) of turfgrasses in the northeastern United States. Pages 421-424 in: Proc. Int. Turfgrass Res. Conf., 4th. R. W. Sheard, ed. Ontario Agricultural College, University of Guelph and the International Turfgrass Society, Guelph, Ontario.

15. Juhnke, M. E., Mathre, D. E., and Sands, D. C. 1984. A selective medium for *Gaeumannomyces graminis* var. *tritici*. Plant Dis. 68:233-236.

16. Landschoot, P. J. 1988. Taxonomy and pathogenicity of ectotrophic fungi with *Phialophora* anamorphs from turfgrasses. Ph.D. dissertation. University of Rhode Island, Kingston.

17. Landschoot, P. J., and Jackson, N. 1989. *Gaeumannomyces incrustans* sp. nov., a root-infecting hyphopodiate fungus from grass roots in the United States. Mycol. Res. 93:55-58.

18. Landschoot, P. J., and Jackson, N. 1989. *Magnaporthe poae* sp. nov., a hyphopodiate fungus with a *Phialophora* anamorph from grass roots in the United States. Mycol. Res. 93:59-62.

19. Lankow, R. K., Woodhead, S. H., Patterson, R. J., Massey, R., and Schochetman, G. 1984. Monoclonal antibody diagnostics in plant disease management. Plant Dis. 68:1100-1101.

20. MacDonald, J. D., and Duniway, J. M. 1979. Use of fluorescent antibodies to study the survival of *Phytophthora megasperma* and *P. cinnamomi* zoospores in soil. Phytopathology 69:436-441.

21. McCarty, L. B., and Lucas, L. T. 1989. *Gaeumannomyces graminis* associated with spring dead spot of bermudagrass in the southeastern United States. Plant Dis. 73:659-661.

22. Miller, S. A., Grothaus, G. D., Petersen, F. P., and Papa, S. L. 1986. Detection of Pythium blight in turfgrass using a monoclonal antibody-based diagnostic test. (Abstr.) Phytopathology 76:1057.

23. Miller, S. A., Grothaus, G. D., Petersen, F. P., Rittenburg, J. H., Plumley, K. A., and Lankow, R. K. 1989. Detection and monitoring of turfgrass pathogens by immunoassay. Pages 109-120 in: Integrated Pest Management for Turfgrass and Ornamentals. A. R. Leslie and R. L. Metcalf, eds. U. S. Environmental Protection Agency, Washington, D. C.

24. Nameth, S. T., Shane, W. W., and Stier, J. C. 1990. Development of a monoclonal antibody for detection of *Leptosphaeria korrae*, the causal agent of necrotic ringspot disease of turfgrass. Phytopathology 80:1208-1211.

25. Phillips, J. M., and Hayman, D. S. 1970. Improved procedures for clearing roots and staining parasitic and vesicular-arbuscular mycorrhizal fungi for rapid assessment of infection. Trans. Brit. Mycol. Soc. 55:158-160.

26. Plumley, K. A., Clarke, B. B., Hillman, B. I., and Bunting, T. E. 1992. The effect of mowing height on the distribution of *Magnaporthe poae* in the soil profile and the development of a DNA probe for the detection of this pathogen. (Abstr.) Phytopathology 82:1160.

27. Sauer, K. M., Hulbert, S. H., and Tisserat, N. A. 1993. Identification of *Ophiosphaerella herpotricha* by cloned DNA probes. Phytopathology 83:97-102.

28. Shane, W. W. 1991. Prospects for early detection of Pythium blight epidemics on turfgrass by antibody-aided monitoring. Plant Dis. 75:921-925.

29. Shane, W. W., and Nameth, S. T. 1988. Monoclonal antibodies for diagnosis of necrotic ring spot of turfgrass. (Abstr.) Phytopathology 78:1521.

30. Smiley, R. W., and Craven Fowler, M. 1984. *Leptosphaeria korrae* and *Phialophora graminicola* associated with Fusarium blight syndrome of *Poa pratensis* in New York. Plant Dis. 68:440-442.

31. Smiley, R. W., Dernoeden, P. H., and Clarke, B. B. 1992. Compendium of Turfgrass Diseases. 2nd ed. American Phytopathological Society, St. Paul, MN.

32. Smiley, R. W., Kane, R. T., and Craven-Fowler, M. C. 1985. Identification of *Gaeumannomyces*-like fungi associated with patch diseases of turfgrasses in North America. Pages 609-618 in: Proc. Int. Turfgrass Res. Conf., 5th. F. Lemaire, ed. INRA Publications and the International Turfgrass Society, Versailles, France.

33. Smith, J. D., Jackson, N., and Woolhouse, A. R. 1989. Fungal Diseases of Amenity Turf Grasses. E. & F. N. Spon, London.

34. Tisserat, N. 1988. Differentiation of *Ophiosphaerella herpotricha* and *Leptosphaeria korrae* by restriction fragment length polymorphism analysis. (Abstr.) Phytopathology 78:1613.

35. Tisserat, N. A., Hulbert, S. H., and Nus, A. 1991. Identification of *Leptosphaeria korrae* by cloned DNA probes. Phytopathology 81:917-921.

36. Tisserat, N. A., Pair, J. C., and Nus, A. 1989. *Ophiosphaerella herpotricha*, a cause of spring dead spot of bermudagrass in Kansas. Plant Dis. 73:933-937.

37. Walker, J. 1980. *Gaeumannomyces, Linocarpon, Ophiobolus* and several other genera of scolecospored ascomycetes and *Phialophora* conidial states, with a note on hyphopodia. Mycotaxon 11:1-129.

38. Walker, J., and Smith, A. M. 1972. *Leptosphaeria narmari* and *L. korrae* spp. nov., two long-spored pathogens of grasses in Australia. Trans. Br. Mycol. Soc. 58:459-466.

39. Worf, G. L., Stewart, J. S., and Avenius, R. C. 1986. Necrotic ring spot disease of turfgrass in Wisconsin. Plant Dis. 70:453-458.

INTEGRATING STRATEGIES FOR THE MANAGEMENT OF PATCH DISEASES CAUSED BY ROOT INVADING ECTOTROPHIC FUNGI

Peter H. Dernoeden
University of Maryland
College Park, MD 20742

INTRODUCTION

Among the most destructive diseases of turfgrasses are those incited by darkly pigmented, ectotrophic, root invading fungi. Currently recognized diseases and pathogens of this group include take-all patch, incited by *Gaeumannomyces graminis* (Sacc.) Arx & D. Olivier var. *avenae* (E. M. Turner) Dennis; summer patch, incited by *Magnaporthe poae* Landschoot & Jackson; necrotic ring spot, incited by *Leptosphaeria korrae* J. C. Walker & A. M. Sm.; bermudagrass decline, associated with *G. graminis* (Sacc.) Arx & D. Olivier var. *graminis*; and spring dead spot. *L. narmari* J. C. Walker & A. M. Sm. is the more common incitant of spring dead spot in Australia (88), whereas *L. korrae* is the documented incitant in California and Maryland (8, 28, 29). *Ophiosphaerella herpotricha* (Fr.:Fr.) J. C. Walker is the causal agent of this disease in Kansas (80, 82), and *G. graminis* var. *graminis* has been associated with spring dead spot in North Carolina (51, 52). Except for bermudagrass decline (see Chapter two), this group of diseases characteristically produces circular patch symptoms. Host range and environmental conditions conducive for pathogen development, however, vary for each disease. The destructive nature of these diseases is due to: (1) the limited availability of disease resistant germplasm; (2) an incomplete understanding of pathogen biology and disease etiology; (3) few field investigations assessing the influence of cultural practices on disease severity; and (4) an inconsistent level of disease control with chemotherapeutants. This chapter will review cultural and chemical practices designed to reduce the severity of patch diseases incited by root pathogens.

TAKE-ALL PATCH

Take-all patch disease, formally known as Ophiobolus patch, is an extremely destructive disease of bentgrass turf in Australia, Europe, and North America (75). The disease also has been reported on annual bluegrass (*Poa annua* L.) in England (74).

Disease Development

The take-all pathogen actively attacks roots during cool, moist weather (32, 62). In the moist climate of western England, where take-all commonly occurs, fewer cases of the disease develop following dry weather in summer and autumn (74). Symptoms of this disease are most conspicuous in the mid-Atlantic region of the United States from late April to early July and in autumn when cool, wet weather prevails (21). Bentgrass infected by *G. graminis* var. *avenae* in the spring may recover by summer; however, if irrigation is withheld in the summer, infected turf may die from drought stress. In the Pacific Northwest, take-all is most evident during late spring and early fall (G. A. Chastagner, *personal communication*).

Soil pH
Take-all patch is common in soils where the pH is between 4.3 and 7.5 (75). The most severe infections occur when soil pH exceeds 7.0 in the upper 2.5 cm of the soil

profile (74). In Maryland, the disease has been most severe in sandy loam soils with a pH of 6.7 to 7.2 (21). Smith (74) was one of the first to report that the disease is more severe following the application of ground limestone. In his studies, finer grades of limestone enhanced disease severity, whereas the use of coarser grades of limestone delayed this response up to three years. The disease did not spread to untreated plots from adjacent plots receiving finer grades of limestone.

Chemical Control

In 1958, Jackson (34) reported that organo-mercury fungicides were more effective than inorganic mercury or cadmium-based fungicides for the control of take-all patch in England. In the United States, Dernoeden et al (22) determined that phenyl mercury acetate (0.3 kg Hg ha^{-1}) and triadimefon (3.0 kg ai ha^{-1}) provided excellent and good disease suppression, respectively, when applied on a curative basis to a Penncross creeping bentgrass (*Agrostis stolonifera* L.) turf. Iprodione (6.1 kg ai ha^{-1}), propiconazole (1.3 kg ai ha^{-1}), and chlorothalonil (12.2 kg ai ha^{-1}) were less effective. Bentgrass foliage became chlorotic with sequential applications of phenyl mercury acetate and exhibited a bluish-purple cast with the application of propiconazole or triadimefon. In a study conducted in western Washington, a single December application of fenarimol (1.5, 3.0, or 6.0 kg ai ha^{-1}), propiconazole (3.0 or 6.0 kg ai ha^{-1}), or triadimefon (3.0 kg ai ha^{-1}) significantly reduced take-all patch the following June (5). The application of fungicides in April, however, was not as effective as a single, December application.

Cultural and Integrated Control

Several researchers have achieved dramatic reductions in disease incidence and severity through the use of various fertilizers, fungicides, and the insecticide chlordane (9, 12, 22, 31, 34, 74). Smith (74) reported excellent control of take-all patch in England when nitrogen in the form of monoammonium phosphate or ammonium sulfate was applied at the rate of 74 kg N ha^{-1}. The fertilizers stimulated plant growth and the subsequent reduction in disease development was attributed to soil acidification. Acidification presumably discouraged the ability of the pathogen to cause disease. In Washington (9), better control of take-all patch was observed with the application of ammonium sulfate plus chlordane or ammonium sulfate plus sulfur-containing compounds when compared with the application of urea plus chlordane. Phosphorus and potassium fertilizers were also reported to reduce the severity of take-all in bentgrass turf (31).

Soil acidification with ammonium forms of nitrogen is the primary cultural means of controlling take-all patch (12). Early studies in the United States used excessively high rates of nitrogen (i.e., 392, 588, and 980 kg N ha^{-1}) to lower soil pH (9, 31). In Maryland, Dernoeden (12) used several acidifying agents at conventional rates and dates of application to suppress disease symptoms. Since phenyl mercury acetate suppressed the severity of take-all patch in earlier studies (22), this fungicide was applied on a preventive basis with either sulfur (90G) or ammonium sulfate to determine if enhanced disease control could be obtained. The materials were applied in the fall of 1983, in the spring and fall of 1984, and in the spring of 1985 (Table 6.1).

In June 1984, patch symptom development was variable, resulting in a large experimental error. Hence, any treatment that reduced the plot area affected by take-all patch to less than one percent was subjectively judged effective in disease control. Using this criterion, ammonium chloride and phenyl mercury acetate plus ammonium sulfate were the two most effective treatments in reducing take-all (Table 6.1). Sulfur, urea, phenyl mercury acetate, and phenyl mercury acetate plus sulfur reduced disease severity, but not to an acceptable level. By June 1985, plots treated with ammonium chloride and ammonium sulfate were free of take-all, and disease activity was reduced to a trace amount in urea-treated plots. Although

Table 6.1. The influence of nitrogen fertilizers, sulfur, and phenyl mercury acetate on take-all patch, thatch production, and pH in a Penncross creeping bentgrass turf.[w]

Treatment[x]	Application rate (kg ha^{-1})	Percent of plot diseased 1984	Percent of plot diseased 1985	Thatch[Y] depth (cm)	pH[Y] Thatch	pH[Y] Soil
Ammonium sulfate	36.6 N	1.2 a[z]	0.0 a	3.2 a	6.6 ab	6.3 b
Ammonium chloride	36.6 N	0.2 a	0.0 a	3.0 abc	6.5 b	6.3 b
Urea	36.6 N	3.7 a	0.3 a	2.9 a-d	6.9 a	6.5 ab
Sulfur 90G	36.6 S	4.7 a	6.0 bc	3.0 abc	6.7 ab	6.4 ab
Sulfur 90G + ammonium sulfate	24.4 S + 24.4 N	1.2 a	3.0 ab	3.1 ab	6.7 ab	6.5 ab
PMA 10L	0.3 Hg	4.7 a	1.3 ab	2.6 cd	6.9 a	6.5 ab
PMA 10L + Sulfur 90G	0.3 Hg + 36.6 S	2.0 a	8.7 c	2.7 bcd	6.7 ab	6.4 ab
PMA 10L + ammonium sulfate	0.3 Hg + 36.6 N	0.3 a	1.0 ab	2.8 a-d	6.9 a	6.4 ab
Untreated control	---	10.7 b	9.0 c	2.5 d	6.9 a	6.6 a

w Dernoeden, P. H., 1987 (12).
x Nitrogen fertilizers and sulfur were applied on a monthly basis in: October to December 1983; April 1984; September to December 1984; and April 1985. Phenyl mercury acetate (PMA) alone or in combination with sulfur or ammonium sulfate was applied monthly in: October and November 1983; April, September and October 1984; and April 1985.
Y Thatch depth was determined in July 1985 and thatch and soil pH were determined in June 1985. For pH measurements, soil was taken from a 2.5 cm zone just below the thatch layer (pH measured in a 1:1 ratio of soil and distilled water).
z Means within a column followed by the same letter are not significantly different ($P < 0.05$) (Bayes LSD).

sulfur did reduce disease severity, it did not provide an acceptable level of control. This finding was not expected and may be due to the inability of granular forms of sulfur to rapidly acidify the soil or thatch (12).

Plots receiving nitrogen or sulfur alone had significantly more thatch than untreated turf when measured by Dernoeden (12) in 1985 (Table 6.1). The long-term influence of thatch depth on the severity of take-all patch is unknown; however, it is likely that management of the thatch microenvironment (i.e., adjusting thatch pH) may be a key factor in reducing the severity of this disease. Due to the buffering capacity of soil and organic matter, the acidification of both thatch and soil in response to fertilization is a slow process (11). The reliance on bulk thatch and soil pH as indicators of disease development, therefore, may be somewhat misleading. This is due to difficulties in accurately assessing the pH of the rhizosphere where the pathogen resides. Soil water adjacent to plant roots often has a much lower pH than is indicated by standard, bulk soil pH tests (66). Acidification of the rhizosphere is believed to be the primary factor responsible for alleviating take-all patch with ammonium forms of nitrogen (66); however, the relationship between the severity of take-all patch and soil pH is not completely understood (12). It is possible that decreasing the rhizosphere pH reduces the growth of *G. graminis* var. *avenae* while favoring the growth of other, possibly beneficial, microorganisms (11).

The use of ammonium chloride as a nitrogen source appears to be the most rapid means of controlling take-all patch without the use of fungicides (12). The chloride anion in ammonium chloride has been shown to lower the water potential of cell sap in wheat roots, thus reducing the ability of the fungus to colonize the host (6). The utility of this compound is in question because it is not readily available and is difficult to handle. Although ammonium sulfate does not appear to reduce take-all patch as rapidly as ammonium chloride, it is an inexpensive and readily available fertilizer (12).

Phosphorus and potassium also have been shown to reduce the severity of take-all patch (31). The use of potassium chloride as a source of potassium should be considered because of a possible link between the chloride anion and disease suppression. The addition of phosphorus, however, is not required in soils that exceed 60 kg P ha^{-1}.

Microbial Antagonism

Several theories have been proposed to explain why take-all patch has appeared recently in the eastern United States. Jackson (35) suggested that mild, chronic disease symptoms have been present for years in bentgrass turf, but that the disease has been either misdiagnosed or dismissed as a cultural problem. Recent restrictions placed on the use of mercury-based fungicides and the current practice of constructing putting greens with high sand content also may be responsible for the recent development of this disease. Sand mixes are often heat-sterilized prior to use and are, therefore, initially devoid of naturally occurring, antagonistic or suppressive microorganisms.

In pot (89) and field (90) studies, *G. graminis* var. *graminis* suppressed take-all patch. Deacon (11) also proposed the use of an antagonist, *Phialophora graminicola* (Deacon) J. Walker, for the control of this disease. The reduction in populations of potential biological control agents in fumigated soils, therefore, may induce severe outbreaks of take-all patch (90). Moreover, the demise of such antagonists in soils where pesticide use has been extensive could also enhance disease severity (11).

Take-all Decline

The severity of take-all patch may diminish within five to seven years after symptoms first appear on bentgrass turf (69, 74). In cereals, symptoms of take-all, caused by *G. graminis* (Sacc.) Arx & D. Olivier var. *tritici* J. Walker and *G. graminis* var. *avenae*, may also diminish after several years in successively cropped stands (75). This may be due to an increase in naturally occurring, antagonistic microorganisms. On a golf course in Brittany, France, Sarniguet and Lucas (61) recently found that the ratio of fluorescent pseudomonads to total bacteria increased from 1:26 outside patch foci to 1:3 within the center of recolonized patches. Moreover, 82% of the fluorescent pseudomonads recovered from recolonized patches were antagonistic to *G. graminis* var. *avenae in vitro*, compared to 12-34% from disease free turf.

Summary

For optimum control of take-all patch, a total of 150 to 200 kg N ha^{-1} from ammonium chloride or ammonium sulfate should be applied to turf annually for at least two years. Ammonium fertilizers provide excellent disease control and improve winter color; however, they also encourage plant growth, and therefore increased mowing of turf into early winter. Phosphorus and potassium (i.e., potassium chloride) should be applied where soil tests indicate a deficiency of either element. The use of topdressing with a pH above 6.0 should be avoided, and thatch should be controlled through aerification and/or verticutting. Where liming is necessary, only the coarsest grades of lime should be used to avoid rapid changes in soil pH. Irrigation water high in pH may also enhance the severity of take-all patch and should, therefore, be monitored.

Currently, only the fungicide fenarimol is labeled for the control of take-all patch on turfgrasses. Phenyl mercury acetate (where legal to apply), propiconazole, or triadimefon may be used in late fall or early winter for the preventive control of snow mold, and this should assist in efforts to reduce the severity of take-all patch. There is little information regarding the relative susceptibility of the many recently marketed creeping bentgrass cultivars. The cultivar Putter, however, is believed to possess improved resistance to take-all patch.

SUMMER PATCH

Summer patch, caused by *M. poae* (46), is an extremely damaging disease of Kentucky bluegrass (*P. pratensis* L.), annual bluegrass, and fine fescue (*Festuca* spp.) turf (65). Until 1984, summer patch was known as Fusarium blight, a disease incited by *Fusarium culmorum* (Wm. G. Sm.) Sacc. and *F. poae* (Peck) Wollenweb. (7, 64). In 1984, Smiley (64) and Smiley and Craven Fowler (67) determined that summer patch was a distinctively different disease from Fusarium blight, and reported that the causal agent was *P. graminicola*. Their isolates were later reclassified in 1987 as *M. poae* by Landschoot and Jackson (46). Smiley and Craven Fowler (67) also reported that *L. korrae* could elicit similar patch symptoms in Kentucky bluegrass, and this disease was later named necrotic ring spot (92). Hence, research conducted prior to 1984 on Fusarium blight is difficult to interpret since any one of these causal agents may have been involved. This has resulted in many of the discrepancies that are discussed in the following sections.

Disease Development

Although environmental and cultural stresses may predispose Kentucky bluegrass to summer patch and Fusarium blight, little is known about the effect of these factors on disease development.

In a field survey conducted by Bean (1), the severity of Fusarium blight increased with the duration and intensity of sunlight. Turf near heat retaining structures or southern facing slopes was often the first to show symptoms of this disease. Fulton et al (30) indicated that 7-14 days of high humidity and warm temperatures ($\geq$ 27 C) were required for symptoms of Fusarium blight to develop. In greenhouse and laboratory studies (70, 71), high temperatures ($\geq$ 29 C) increased the ability of *M. poae* (reported as *P. graminicola*) to grow and cause disease. Kackley et al (38) observed that *M. poae*

131

attained maximum growth on potato dextrose agar and was most destructive to inoculated Kentucky bluegrass turf at 25-30 C.

Cultural Control

Soil Factors

Soil factors play an important role in the development of summer patch. Summer patch can be more severe on poorly drained and compacted sites. As a result, aerification is often required to reduce soil bulk density, increase rooting, and reduce symptom expression (B. B. Clarke, *personal communication*).

Another factor that affects the incidence of summer patch is the availability of soil moisture. Literature describing the effect of soil moisture on Fusarium blight prior to its separation from summer patch and necrotic ring spot, however, is often contradictory. In 1969, Bean (2) observed that Kentucky bluegrass swards were most severely damaged by Fusarium blight when either drought stressed or maintained under light, frequent watering. Endo and Colbaugh (27) also reported that Fusarium blight was more severe where Kentucky bluegrass was subjected to temporary drought stress. Fulton et al (30), however, reported in 1974 that Fusarium blight was most severe following warm and wet periods.

Before *M. poae* or *P. graminicola* were associated with the Fusarium blight complex, Smiley (63) observed that major outbreaks of Fusarium blight on Long Island, New York were associated with periods of abundant moisture or with alternating periods of wetness and drought. Disease outbreaks were especially severe when hot weather ($\geq$ 30 C) followed a period of heavy rainfall. He later determined that *M. poae* (reported as *P. graminicola*) grew most rapidly *in vitro* at high water potentials (-0.1 MPa) (71). Kackley et al (38) observed that disease severity induced by *M. poae* was greatest on Kentucky bluegrass grown at soil matric potentials between -0.12 and -0.35 MPa and at temperatures between 25 and 30 C. In field studies, Kackley et al (37) showed that summer patch was more severe in non drought stressed (> -0.05 MPa) than in drought stressed (< -0.05 MPa) Kentucky bluegrass turf.

Mowing Height

Turgeon (84) stated that close mowing produces a turf that is aesthetically pleasing but often less tolerant of environmental stress and disease pressure. In a field study, Turgeon and Meyer (85) found that mowing height affected the susceptibility of Merion, Nugget, and Pennstar, but not Fylking or Kenblue Kentucky bluegrasses, to Fusarium blight. Disease severity on Pennstar turf was less at a mowing height of 3.8 cm than at a height of 1.9 cm; conversely, disease severity on Merion and Nugget turf was less at a mowing height of 1.9 cm.

Recent studies conducted with summer patch support the concept that disease severity is reduced at higher mowing heights. Davis and Dernoeden (10) reported that S-21 Kentucky bluegrass turf maintained at 7.6 cm was damaged less by summer patch, exhibited improved quality, and contained higher levels of total non structural carbohydrates than turf maintained at 3.8 cm. Plumley et al (57), however, found no correlation between the height of cut and the movement of *M. poae* through the soil profile in a Baron Kentucky bluegrass turf maintained at 4.0 and 8.0 cm. In a greenhouse study, unmowed Merion Kentucky bluegrass turf sustained less damage from *M. poae* (reported as *P. graminicola*) than turf mowed at 2.0 cm (71).

Nitrogen Fertility

High nitrogen fertility levels are generally believed to enhance the severity of Fusarium blight and summer patch. In an early report, Bean (2) was unable to associate Fusarium blight with high nitrogen levels. Later, Turgeon and Meyer (85) observed that the severity of Fusarium blight in the field varied with Kentucky bluegrass cultivar, mowing height, and nitrogen fertility level. In all cultivars tested except Kenblue, high spring fertilization increased the severity of Fusarium blight. Nash (55) evaluated various nitrogen sources on summer patch development in Merion Kentucky bluegrass. He observed that summer patch was more severe in plots treated with urea than with slow release nitrogen sources (e.g., methylene urea, activated sewage sludge, and sulfur coated urea).

The influence of nitrogen source, irrigation, and mowing height on the severity of summer patch on S-21 Kentucky bluegrass turf was assessed in a comprehensive field study at the University of Maryland (10). Ammonium chloride (a readily available, acid reacting nitrogen source), sulfur coated urea (a slow release, acid reacting nitrogen source), sodium nitrate (a readily available, neutral reacting nitrogen source), and urea (a readily available, slowly acidifying nitrogen source) were selected for their ability to affect soil pH. The parameters, as noted in Tables 6.2 and 6.3, were arranged in factorial design.

The severity of summer patch was reduced under high (7.6 cm) mowing practices in both years of the study. Symptoms were least apparent on turf receiving deep, infrequent irrigation and maintained at a mowing height of 7.6 cm (Table 6.2). Disease severity was greatest on turf receiving light, frequent irrigation and maintained at a mowing height of 3.8 cm (Table 6.2). Of the nitrogen treatments evaluated, the severity of summer patch was lowest in those plots fertilized with sulfur coated urea and greatest in those plots fertilized with sodium nitrate (Table 6.3). The authors speculated that the benefits of using sulfur coated urea may be due to the acidifying and slow release properties of this nitrogen source. The authors also hypothesized that disease severity increased with sodium nitrate because this nitrogen source reduced rooting and did not modify soil pH.

In the first year of a study conducted by Thompson et al (78), ammonium sulfate was found to slightly reduce the severity of summer patch in Fylking Kentucky bluegrass. During the second year of the study, ammonium sulfate significantly reduced both the onset and severity of this disease

134

Table 6.2. The interaction between irrigation and mowing height on the development of summer patch (*M. poae*) in S-21 Kentucky bluegrass.[v]

Irrigation Treatment[w]	Mowing height[x] (cm)	Plot area damaged, 1987 (%)				
		17 Jul	20 Aug	30 Aug	10 Sept	17 Sept
Light frequent	3.8	17 a[y]	36 b	24 a	26 ab	24 b
Light frequent	7.6	6 a	16 d	14 b	13 d	14 c
Light frequent	7.6 recovery[z]	---	37 ab	26 a	23 bc	19 bc
Deep infrequent	3.8	19 a	38 a	28 a	29 a	31 a
Deep infrequent	7.6	0 a	4 e	2 c	3 e	1 d
Deep infrequent	7.6 recovery	---	33 c	23 a	17 cd	16 c

[v] Davis, D. B., and Dernoeden, P. H., 1991 (10).

[w] Irrigation treatments were initiated May 1987. Plots receiving light frequent irrigation were watered to a depth of 1.5 to 2.0 cm. Plots receiving deep infrequent irrigation were watered to a depth of 15 to 20 cm.

[x] Mowing treatments were initiated in March 1987 and the plots were mowed on an as-needed basis.

[y] Means in a column followed by the same letter are not significantly different ($P < 0.05$) (Bayes LSD).

[z] In mid-August 1987, recovery plots were allowed to regrow from 3.8 to 7.6 cm. The height of cut was not lowered to 3.8 cm until mid-April 1988.

135

Table 6.3. Effect of four nitrogen sources on the development of summer patch (*M. poae*) in S-21 Kentucky bluegrass.[x]

Nitrogen source[Y]	Plot area damaged, 1987							
	17 Jul	21 Jul	31 Jul	4 Aug	13 Aug	20 Aug	30 Aug	10 Sept
	(%)							
Urea	10 b[z]	12 ab	16 a	24 ab	32 a	27 b	22 a	20 a
Sulfur coated urea	6 b	7 b	8 b	16 b	22 b	20 c	14 b	12 b
Ammonium chloride	10 b	12 ab	15 ab	24 ab	30 a	28 ab	20 a	20 a
Sodium nitrate	16 a	18 a	22 a	31 a	35 a	34 a	23 a	23 a

[x] Davis, D. B., and Dernoeden, P. H., 1991 (10).

[Y] All fertilizers were applied at the rate of 49 kg N ha^{-1} in September, October, and November 1986 and May 1987.

[z] Means in a column followed by the same letter are not significantly different ($P < 0.05$) (Bayes LSD).

when compared to turf treated with calcium nitrate or left unfertilized. The authors attributed this effect primarily to a reduction in rhizosphere pH (78). In a related study, ammonium chloride, ammonium sulfate, and sulfur coated urea reduced disease severity whereas calcium nitrate and potassium nitrate intensified disease symptoms in comparison to non fertilized turf (D. C. Thompson, *personal communication*). Thompson and Clarke (77) were also able to reduce the severity of summer patch 25-36% compared to controls in field grown Baron Kentucky bluegrass using antagonistic bacteria recovered from a take-all suppressive soil. Further research is needed, however, to assess the potential of using antagonistic microorganisms to control summer patch in the field.

Disease Resistant Grasses

Due in large part to the destructive nature of summer patch, most Kentucky bluegrass golf course fairways in the Baltimore-Washington, D.C. corridor have been replaced by perennial ryegrass (*Lolium perenne* L.). Perennial ryegrass, tall fescue (*Festuca arundinacea* Schreb.), creeping bentgrass, and warm-season grasses appear to be resistant, or perhaps immune, to summer patch in the field (15).

The primary species affected by *M. poae* are Kentucky bluegrass, annual bluegrass, and fine fescues (15, 65). In Maryland, Pennlawn and Flyer strong creeping red fescue (*F. rubra* subsp. *rubra* L.) are considered susceptible to summer patch, whereas Aurora hard fescue (*F. longifolia* Thuill.) is moderately resistant. According to Kemp (39) and Kemp et al (40, 41), strong creeping red fescues are less susceptible than hard fescues or slender creeping red fescues (*F. rubra* subsp. *litoralis* (Meyer) Auquier) in naturally infected stands maintained at 3.8 cm in New Jersey. The authors also report that Chewings fescues (*F. rubra* subsp. *commutata* Gaud.) are moderately susceptible to this disease.

Among Kentucky bluegrass cultivars, Adelphi, Aspen, Enmundi, Rugby, Sydsport, and Touchdown appear to have good summer patch resistance, whereas Dormie, Fylking, S-21, Merion, and Windsor are very susceptible (68). The relative susceptibility of Kentucky bluegrass cultivars to summer patch may vary due to complex soil, climatic, and cultural factors that affect root infection and disease development. Varietal susceptibility may be further complicated by the coexistence of *M. poae* with other root-infecting fungi, such as *L. korrae* and *G. incrustans* Landschoot & Jackson (39).

Chemical Control

Control of summer patch with fungicides is often erratic and expensive. Fungicide efficacy is influenced by timing, rate, and amount of water applied with the fungicide (15). Smiley (63) extensively reviewed the use and efficacy of fungicides for the control of Fusarium blight. Benomyl, fenarimol, propiconazole, thiophanate-methyl, and triadimefon are among the fungicides most commonly used for the control of summer patch (15). Propiconazole, fenarimol, and triadimefon applied in 800-1000 L H_2O ha^{-1} prior to symptom expression generally provide better control of summer patch than the benzimidazole fungicides benomyl or thiophanate-methyl (13, 19, 20, 33, 43, 58, 63). Benzimidazole fungicides, however, may significantly reduce disease severity when applied after symptoms begin to appear (19, 20). Landschoot and Clarke (43) reported that the efficacy of benzimidazole fungicides increased when applied in 2000 or 4000 L H_2O ha^{-1} compared to 800 L H_2O ha^{-1}. Cyproconazole, myclobutanil, and tebuconazole are examples of experimental fungicides that have provided excellent control of this disease in fungicide-amended media (76). The use of chlorothalonil (9.2 kg ai ha^{-1}) (18) and iprodione (6.1 kg ai ha^{-1}) (B. B. Clarke, *personal communication*) have been shown to enhance the severity of summer patch in certain Kentucky bluegrass cultivars. These contact fungicides, therefore, should not be used in areas prone to this disease.

Summary

To reduce the severity of summer patch in Kentucky bluegrass or creeping red fescue lawns, maintain mowing height at 6-8 cm, water deeply (> 10 cm) and infrequently (only at the onset of drought stress), use slow release nitrogen fertilizers, reduce compaction through aerification, and overseed with resistant Kentucky bluegrass or fine fescue cultivars. When renovating, plant turfgrass species that are resistant to summer patch (i.e., bentgrass, perennial ryegrass, tall fescue, or resistant Kentucky bluegrass cultivars). For preventive control of summer patch, fenarimol, propiconazole, or triadimefon are generally most effective when applied 30 days prior to symptom development. Curative applications of benomyl or thiophanate-methyl in large volumes of water (2000-4000 L H_2O ha^{-1}) should improve efficacy and provide a satisfactory level of control under most conditions. Multiple applications of these fungicides, applied at 21-28 day intervals, appear to be the most effective means of reducing summer patch in annual bluegrass maintained under putting green conditions. Although many fungicides reduce the severity of summer patch, the level of control by any particular fungicide may vary from year to year. Moreover, the level of disease control may be commercially unacceptable if disease pressure is extremely high.

SPRING DEAD SPOT

Spring dead spot is the most destructive disease of bermudagrass (*Cynodon dactylon* (L.) Pers.) and bermudagrass hybrids in North America and Australia (69). Although the etiology of this disease has been mentioned in previous chapters, a discussion of this information follows so that the association between specific causal agents and disease control can be clearly understood.

In 1960, spring dead spot was first described in Oklahoma by Wadsworth and Young (87). To date, several root-infecting fungi have been associated with this disease in the United States. The incitant of spring dead spot has been identified in California and Maryland as *L. korrae* (8, 29); in Kansas as *O. herpotricha* (80, 82); and in North Carolina as *G. graminis* var. *graminis* (51, 52). The causal agents of spring dead spot in Australia have been identified as *L. narmari* and *L. korrae* (72, 88).

Disease Development

Spring dead spot usually occurs where the weather is sufficiently cold in the winter to induce a long dormancy in bermudagrass turf (48, 86). This disease is typically associated with turfs that are three or more years old (56). Lucas (48) reported that spring dead spot occurs in the United States from Maryland to North Carolina, westward through Texas and Oklahoma, and in California (Fig. 2.3). These regions correspond to areas where the average daily temperature in November is below 13 C (48). Spring dead spot has not been documented in the southernmost regions of the United States; however, it was observed at the Louisiana State University Experiment Station turf plots in Baton Rouge, Louisiana during the spring of 1989 (N. Jackson, *personal communication*).

Disease Transmission
Pair et al (56) reported that spring dead spot can be transmitted under field conditions. In their study, plots of 24 bermudagrass clones were inoculated in the fall with naturally infected Midway bermudagrass plugs (causal agent undetermined). Transmission was verified for all clones within two to four years following inoculation. In Maryland, early summer sprigging with stolons of Tufcote bermudagrass infected with *L. korrae* resulted in the appearance of mild disease symptoms the following spring (16).

Cultural Control

Little is known about the influence of cultural practices on the severity of spring dead spot. Although this disease is generally associated with mature, intensively managed turfs (69, 87), severe outbreaks have been observed in Tufcote bermudagrass grown under low management (i.e., where turf received no supplemental irrigation, little or no nitrogen fertilizer, and where weeds were not controlled) (16).

Soil and Management Factors

Factors such as soil texture, compaction, and pH have been reported to affect the incidence and severity of spring dead spot. Young et al (93) observed that soils high in clay content (15-22% clay) were more conducive to disease development than lighter soils (12.5-12.6% clay). Pair et al (56) also found that disease incidence was greater in clay soils. In a field study conducted in Maryland (16), there was a negative association between soil pH and spring green-up of Tufcote bermudagrass turf, and a positive correlation between soil pH and the severity of spring dead spot. Other factors that favor disease development include low mowing height, thatch accumulation, and soil compaction (48).

Nitrogen Fertility

Spring dead spot is usually more severe on turfs maintained at high fertility. High rates of nitrogen applied in late summer (48) or late fall (53) have been shown to enhance symptom expression. For example, Lucas (47) reported that ammonium nitrate (49 kg N ha^{-1}) enhanced disease severity (causal agent unknown) in North Carolina when applied in August and September.

The use of potassium fertilizer to improve winter hardiness of bermudagrass is well documented (36, 48). Until recently, the influence of nitrogen and potassium source on the incidence of spring dead spot had not been addressed. In a greenhouse and growth chamber study, applications of potassium chloride or ammonium sulfate increased the survival of Tufcote bermudagrass plants inoculated with *L. korrae* by 24% when compared to unfertilized, inoculated plants (16). The combination of potassium chloride and ammonium sulfate, however, did not consistently increase percent survival. The application of sodium nitrate reduced the survival of inoculated plants by 32%.

In the same study, Dernoeden et al (16) also evaluated the effect of nitrogen and potassium alone (49 kg N or K ha^{-1}) or in combination on the severity of spring dead spot at two field sites. In general, ammonium sulfate, ammonium chloride, ammonium sulfate plus potassium chloride, or ammonium chloride plus potassium chloride enhanced spring green-up and reduced disease severity by either the second or third year of the study (Table 6.4). In contrast to the greenhouse and growth chamber study, potassium chloride alone did not have a major affect on disease severity; however, trends at both sites suggested that the use of potassium chloride was agronomically important. By mid-summer, all turf treated with nitrogen had significantly recovered from spring dead spot (caused by *L. korrae*) when compared to unfertilized turf. The study demonstrated that acidification reduced disease severity over a two to three year period.

Disease Resistant Germplasm

By 1960, spring dead spot had only been observed on the bermudagrass varieties African, U-3, Common, Tiffine, and Tifgreen (87). At present, of the turf-type bermudagrass cultivars grown in Maryland and northern Virginia, only Tufcote is highly susceptible to spring dead spot. The cultivar Vamont is slightly susceptible to this disease, whereas Midiron and Common are rarely affected. According to Lucas, however, Common, Midiron, and Vamont can be severely affected by spring dead spot in the Piedmont regions of North Carolina (L. T. Lucas, *personal communication*). The bermudagrass cultivars Tifgreen, Common, and Santa Ana have

Table 6.4. Effect of nine fertilizer treatments on spring green-up and the severity of spring dead spot in Tufcote bermudagrass at Silver Spring, MD.[v]

Fertilizer[w]	Plot area damaged, 1989[x] (%)				Green-up[x] (%)		Soil pH[Y]
	25 May	6 Jun	23 Jun	6 Jul	1989	1990	
NaNO$_3$	44 ab[z]	39 ab	11 bc	6 ab	22 abc	26 cde	6.3 a
(NH$_4$)$_2$SO$_2$	32 b	31 ab	8 bcd	5 b	25 ab	51 ab	4.6 f
NH$_4$Cl	32 b	30 ab	5 cd	5 b	20 abc	49 abc	4.8 f
Urea	59 a	46 a	14 ab	5 b	14 c	23 de	5.6 d
KCl	54 ab	41 ab	16 ab	10 a	14 c	40 a-d	5.9 bcd
NaNO$_3$ + KCl	46 ab	40 ab	10 bcd	4 b	19 abc	14 e	6.2 ab
(NH$_4$)$_2$SO$_2$ + KCl	43 ab	25 b	3 d	2 b	26 ab	59 a	4.5 f
NH$_4$Cl + KCl	34 b	26 b	2 d	2 b	29 a	56 a	5.2 e
Urea + KCl	56 ab	42 ab	9 bcd	4 b	16 bc	16 de	5.8 cd
No fertilizer	58 a	46 a	20 a	10 a	16 bc	32 b-e	6.1 abc

[v] Dernoeden et al, 1991 (16).
[w] Fertilizers were applied on 27 June, 22 July, and 15 August in 1986; 20 May, 20 June, 2 August, and 6 September in 1987; 1 June, 1 July, 1 August, and 2 September in 1988; and 12 May, 12 June, and 11 July in 1989.
[x] Percent green-up (assessed in May, 1989 and 1990) and percent disease were visually assessed on a 0 to 100% scale, where 0 = entire area brown or dead, and 100 = entire plot area green and healthy.
[Y] Soil pH was assessed on samples collected 10 May 1989.
[z] Means in a column followed by the same letter are not significantly different (P < 0.05) (DMRT).

143

been reported susceptible to this disease in California (28). Although differences in susceptibility among bermudagrass cultivars have been observed, efforts to develop highly disease resistant germplasm have been largely unsuccessful. Bermudagrass cultivars that have a high degree of winter-hardiness, however, appear to be less susceptible to spring dead spot (75).

Chemical Control

The control of spring dead spot with fungicides has yielded varying results. This is not surprising since this disease is caused by different fungi in different regions, and that in earlier work, the causal agents were not identified. To clarify these discrepancies, information on chemical control is presented by causal agent.

Early Work: Etiology Unknown

In 1974, Kozelnicky reported that nabam applied four times in the fall at monthly intervals controlled spring dead spot in Missouri (42). Lucas, however, reported that nabam, carboxin, chloroneb, and maneb were not effective when applied on Tifton 419 bermudagrass in North Carolina (47). In that study, satisfactory control was obtained with five monthly applications (July to November) of benomyl (1.56 g ai m^{-2}) or with a combination of benomyl (0.77 g ai m^{-2}) plus chlorothalonil (0.97 g ai m^{-2}), PCNB (1.52 g ai m^{-2}), and chloroneb (0.97 g ai m^{-2}). In two companion studies, Lucas (47) determined that the critical application time for the control of spring dead spot with benomyl is October or November, before winter dormancy of bermudagrass occurs.

Leptosphaeria spp.

Smith (73) reported that nabam or thiram applied at four week intervals from late summer to early spring controlled *Leptosphaeria* spp. in Australia. Dernoeden (14) observed that fenarimol (6.1 kg ai ha^{-1}) applied in October 1982 and in

September and October 1983 to Tufcote bermudagrass plots controlled spring dead spot caused by *L. korrae* in May 1986. In another study (17), fenarimol (6.1 kg ai ha⁻¹), tebuconazole (3.2 kg ai ha⁻¹), or diniconazole (3.2 kg ai ha⁻¹) applied twice in September to Tufcote bermudagrass reduced disease severity the following spring.

In several studies conducted in Maryland (P. H. Dernoeden, *unpublished data*), single mid-September applications of tebuconazole (3.1 kg ai ha⁻¹), benomyl (12.2 kg ai ha⁻¹), or fenarimol (1.5 kg ai ha⁻¹) reduced the severity of spring dead spot in Tufcote bermudagrass the following spring. Although these fungicide treatments reduced disease severity, they did not always provide commercially acceptable levels of control.

G. graminis var. *graminis*

In a 1987 North Carolina study, McCarty et al (54) reported that a single application of benomyl (12.2 kg ai ha⁻¹) in September, fenarimol (1.5 or 2.2 kg ai ha⁻¹) in September or October, propiconazole (2.5 kg ai ha⁻¹) in September, or thiophanate-methyl (12.2 kg ai ha⁻¹) in October reduced the severity of spring dead spot in Tifway bermudagrass. Lucas and Newnam (49) later reported that a single September application of fenarimol (1.5 kg ai ha⁻¹), or single October applications of propiconazole (2.5 kg ai ha⁻¹), benomyl (12.2 kg ai ha⁻¹), thiophanate-methyl (12.2 kg ai ha⁻¹), or diniconazole (3.0 kg ai ha⁻¹) in Tifway bermudagrass also provided satisfactory disease control. Regrowth of Tifway bermudagrass in a diseased fairway was improved when fenarimol (1.5 and 3.0 kg ai ha⁻¹) was applied in September or October (50).

O. herpotricha

Control of *O. herpotricha* with fungicides has been largely unsuccessful. Tisserat et al (79, 81) evaluated the efficacy of selected fungicides for the control of spring dead spot in a naturally infected Kansas Improved bermudagrass lawn. Although none of the fungicides tested significantly reduced disease severity, fenarimol (3.0 kg ai ha^{-1}) and propiconazole (0.8 kg ai ha^{-1}) tended to reduce the number of infection centers per plot. Benomyl (9.2 kg ai ha^{-1}), however, increased disease severity in both studies. Dernoeden (14) also observed an increase in disease severity when benomyl (6.1 kg ai ha^{-1}) was applied three times to Tufcote bermudagrass infected with *L. korrae*.

In 1991, Tisserat et al (83) reported that fenarimol (0.75 and 1.5 kg ai ha^{-1}) and propiconazole (0.8 and 1.6 kg ai ha^{-1}) reduced disease incidence and severity in artificially inoculated A-29 bermudagrass turf. The level of control achieved with these fungicides, however, was commercially unacceptable.

Summary

Cultural practices can help reduce the incidence and severity of spring dead spot. Since the disease is generally associated with management practices that delay fall dormancy (47, 56), late summer or fall applications of nitrogen fertilizers should be avoided. Ammonium fertilizers applied monthly at spring green-up, with or without potassium chloride, may reduce disease severity and speed recovery of infected turf. Benomyl, fenarimol, propiconazole, tebuconazole, or thiophanate-methyl are most effective when applied in late fall. Although increasing the volume of water applied with systemic fungicides has improved their efficacy with other patch diseases (43), this approach has not been assessed with spring dead spot.

Weeds invading turf areas affected by spring dead spot should be controlled to reduce competition with bermudagrass. The removal of excess thatch may also alleviate disease severity. Selection of bermudagrass cultivars with good winter hardiness should help reduce the severity of spring dead spot. Slow release nitrogen fertilizers, bioorganic amendments, and other cultural practices need to be evaluated to determine their effects on this disease.

NECROTIC RING SPOT

Necrotic ring spot, caused by *L. korrae*, is a serious disease of Kentucky bluegrass turf in North America (4). In 1983, necrotic ringspot was identified as a separate component in the Fusarium blight complex (91). To date, necrotic ring spot has been observed primarily on Kentucky bluegrass turf grown in the northeast, upper midwest, and Pacific northwest regions of the United States (15) (Fig. 2.4).

Disease Development

Environmental conditions that favor the development of necrotic ring spot are similar to those that favor take-all patch (75). Symptoms can occur throughout the growing season during cool, wet weather. Depending on the geographic region in which it develops, necrotic ring spot may appear from late spring to early autumn. Patches often fade with the advent of warmer temperatures in the summer, but frequently reappear in response to heat and drought stress (92). Infection centers may develop again in late autumn when root growth is reduced and persist until early spring (75). The disease is most common on sodded lawns, but may occur on seeded sites, within two to eight years of establishment (4, 92). As with summer patch and take-all patch, disease severity often subsides after several years (75, 91).

Cultural Control

Little is known regarding the cultural control of necrotic ring spot. Drought stress may play a more important role in the development of this disease than summer patch. For example, researchers at Michigan State University observed that light, daily irrigation between May and September reduced the incidence of necrotic ring spot. Irrigation applied biweekly also limited disease development when compared to plots receiving only natural rainfall (J. Vargas, *personal communication*).

The severity of necrotic ring spot has been shown to be reduced when slow-release nitrogen sources are used to fertilize turf (69). In Michigan, bioorganic fertilizers applied between April and September reduced disease incidence in Kentucky bluegrass (J. Vargas, *personal communication*). In a similar field study in Pennsylvania, however, several organic and synthetic forms of nitrogen did not affect the severity of this disease over a period of two years (44). Necrotic ring spot can occur over a wide range of soil pH (5.0-8.0) and is intensified by soil compaction (69).

Disease Resistant Germplasm

In Wisconsin, 22 cultivars of Kentucky bluegrass were evaluated for susceptibility to necrotic ring spot. None of the cultivars tested were highly resistant to this disease, although some were clearly more tolerant than others. Only a slight to moderate level of disease resistance was observed in Midnight, Wabash, Eclipse, Adelphi, Park, and I-13. The cultivars Baron, Birka, Columbia, Georgetown, Glade, Haga, Nassau, Newport, Ram I, Sydsport, and Trampas were very susceptible. When sod growers in Wisconsin changed their seed mixtures containing Ram I, Glade, Sydsport, and Baron to blends containing Midnight, Eclipse, and Adelphi, their problems with necrotic ring spot greatly diminished (G. L. Worf, *personal communication*).

In New York state, Merion, Touchdown, and A-34 Kentucky bluegrass were reported as more susceptible to necrotic ring spot than Adelphi, A-20, or H-7 (68). In a similar evaluation of Kentucky bluegrass varieties and selections in Pennsylvania, Landschoot and Hoyland (45) found that America, Adelphi, Cynthia, Eclipse, I-13, Mystic, Somerset, and Wabash were also highly resistant to this disease.

Chemical Control

The frequent appearance of both necrotic ring spot and summer patch during the summer months has contributed to confusion surrounding the diagnosis and control of these diseases with fungicides. In culture, triadimefon and iprodione were less effective in suppressing the growth of *L. korrae* than propiconazole, fenarimol, phenyl mercury acetate, or benomyl (3, 67). In a field study, Chastagner and co-workers (3, 4) reported that fenarimol, propiconazole, diniconazole, or thiophanate-methyl applied once in the spring controlled necrotic ring spot through the fall. Long-term control of this disease, however, was only attained when fungicides were reapplied each year. Sanders and Soika (59, 60), however, obtained good control of necrotic ring spot with five monthly applications (April through August) of triadimefon (3.2 kg ai ha^{-1}), fenarimol (0.75 kg ai ha^{-1}), or iprodione (6.6 kg ai ha^{-1}). Propiconazole (1.6 kg ai ha^{-1}) and benomyl (12.2 kg ai ha^{-1}) were not effective. The disease also was controlled with myclobutanil (1.6 kg ai ha^{-1}) and cyproconazole (0.81 kg ai ha^{-1}). The authors noted that the control of necrotic ring spot with triadimefon under field conditions was surprising, since isolates of *L. korrae* from the test were not suppressed *in vitro* by this fungicide (60).

Summary

Effective management of necrotic ring spot includes judicious irrigation to prevent drought stress, the use of slow-release nitrogen fertilizers, and overseeding with resistant cultivars of Kentucky bluegrass. When renovating, plant turfgrass species that are resistant to necrotic ring spot (i.e., bentgrass, perennial ryegrass, tall fescue, or resistant Kentucky bluegrass cultivars). Early spring application of fenarimol, propiconazole, or thiophanate-methyl may reduce the incidence and severity of necrotic ring spot when applied on a preventive basis.

BERMUDAGRASS DECLINE

Bermudagrass decline is a destructive root disease of bermudagrass cultivars and hybrids (*C. dactylon* X *C. transvaalensis* Burtt-Davy) managed as golf course putting greens (23). *G. graminis* var. *graminis* has recently been implicated as the primary incitant of this disease. Other fungi such as *G. incrustans* and *Phialophora* spp. have been isolated from bermudagrass exhibiting symptoms of decline; however, their pathogenicity has not been established (23, 26).

Disease Development

In the southeastern United Sates, bermudagrass decline is most often observed during the summer and autumn when hot, humid weather and high amounts of rainfall are common (23). In southern Florida, the decline is apparent during periods of daily rainfall and when the average daily temperature and relative humidity remain fairly constant ($\geq$ 27 C and 75%, respectively) (23). Symptom expression of bermudagrass decline appears to be associated with mowing height. To date, only closely-mowed putting greens have been observed with this disease. The pathogen, however, has been isolated from asymptomatic turf in higher-mowed areas adjacent to infected greens or on lawns and sports fields (69).

Ectotrophic fungi such as *G. graminis* var. *graminis* have been found in association with the roots of bermudagrass sprigs used to establish greens. Since greens are often established on fumigated soil, populations of organisms suppressive to root-infecting fungi are typically absent or present at low levels. In the absence of suppressive microorganisms, root-infecting fungi introduced with bermudagrass sprigs may spread rapidly to new roots. This may explain why putting greens only two and three years old are often affected by this disease (23).

Cultural and Chemical Control

Since the principal incitant of bermudagrass decline was first described in 1991 (23), little information concerning cultural and chemical management practices exists. In a recent study conducted by Elliott (24), several cultural and chemical control measures were evaluated for their ability to control bermudagrass decline on a golf course putting green planted with Tifgreen 328 bermudagrass. The application of fenarimol, propiconazole, triadimefon, or thiophanate-methyl (rates not stated) on a curative basis appeared to inhibit the recovery of turf infected with *G. graminis* var. *graminis*. Plots receiving light topdressing and an increased mowing height, however, had significantly higher quality scores than the other treatments in the test.

Summary

It is anticipated that the control of bermudagrass decline may be best achieved through practices that enhance root development and alleviate plant stress. Elliott and Freeman (25) have suggested that practices used to manage spring dead spot may be helpful in reducing the severity of bermudagrass decline. Mowing height should be raised to avoid plant stress and acidifying fertilizers such as ammonium sulfate should be used to maintain adequate soil fertility. The use of hydrated lime to control algae should be avoided. Core cultivation followed by topdressing with a pathogen-free soil mixture should be used to reduce compaction and improve drainage. Research is currently underway to further evaluate fungicides for the control this disease.

ACKNOWLEDGMENTS

I am grateful to the following individuals for providing unpublished data and other information presented in this review: Dr. Gary A. Chastagner, Dr. Bruce B. Clarke, Dr. C.

Reed Funk, Dr. Bradley I. Hillman, Dr. Noel Jackson, Dr. Leon T. Lucas, Prof. Patricia L. Sanders, Dr. Joseph M. Vargas Jr., and Dr. Gayle L. Worf.

LITERATURE CITED

1. Bean, G. A. 1966. Observations on Fusarium blight of turfgrasses. Plant Dis. Rep. 50:942-945.
2. Bean, G. A. 1969. The role of moisture and crop debris in the development of Fusarium blight of Kentucky bluegrass. Phytopathology 59:479-481.
3. Chastagner, G. A., Goss, R. L., Staley, J. M., and Hammer, W. 1984. A new disease of bluegrass turf and its control in the Pacific Northwest. (Abstr.) Phytopathology 74:811-812.
4. Chastagner, G. A., and Hammer, B. 1987. Current research on necrotic ring spot. Pages 94-95 in: Proc. Northwest Turfgrass Conf., 41st. Northwest Turfgrass Association, Gleneden Beach, OR.
5. Chastagner, G. A., and Staley, J. M. 1989. Fungicidal control of take-all patch. (Abstr.) Phytopathology 79:1169.
6. Christensen, N. W., Taylor, R. G., Jackson, T. L., and Mitchell, B. L. 1981. Chloride effects on water potentials and yield of winter wheat infected with take-all root rot. Agron. J. 73:1053-1058.
7. Couch, H. B., and Bedford, E. R. 1966. Fusarium blight of turfgrasses. Phytopathology 56:781-786.
8. Crahay, J. N., Dernoeden, P. H., and O'Neill, N. R. 1988. Growth and pathogenicity of *Leptosphaeria korrae* in bermudagrass. Plant Dis. 72:945-949.
9. Davidson, R. M., Jr., and Goss, R. L. 1972. Effects of P, S, N, lime, chlordane, and fungicides on Ophiobolus patch disease of turf. Plant Dis. Rep. 56:565-567.
10. Davis, D. B., and Dernoeden, P. H. 1991. Summer patch and Kentucky bluegrass quality as influenced by cultural practices. Agron. J. 83:670-677.

11. Deacon, J. W. 1973. Factors affecting occurrence of the Ophiobolus patch disease of turf and its control by *Phialophora radicicola*. Plant Pathol. 22:149-155.

12. Dernoeden, P. H. 1987. Management of take-all patch of creeping bentgrass with nitrogen, sulfur, and phenyl mercury acetate. Plant Dis. 71:226-229.

13. Dernoeden, P. H. 1989. Preventative control of summer patch in red fescue turf, 1988. Fungic. Nematic. Tests 44:255.

14. Dernoeden, P. H. 1989. Reduced spring dead spot damage with Rubigan two years after application, 1986. Fungic. Nematic. Tests 44:242.

15. Dernoeden, P. H. 1989. Symptomatology and management of common turfgrass diseases in transition zone and northern regions. Pages 273-296 in: Integrated Pest Management for Turfgrass and Ornamentals. A. R. Leslie and R. L. Metcalf, eds. U. S. Environmental Protection Agency, Washington, D. C.

16. Dernoeden, P. H., Crahay, J. N., and Davis, D. B. 1991. Spring dead spot and bermudagrass quality as influenced by nitrogen source and potassium. Crop Sci. 31:1674-1680.

17. Dernoeden, P. H., and McHenry, J. 1989. Preventative control of spring dead spot, 1987. Fungic. Nematic. Tests 44:241.

18. Dernoeden, P. H., and McIntosh, M. S. 1991. Disease enhancement and Kentucky bluegrass quality as influenced by fungicides. Agron. J. 83:322-326.

19. Dernoeden, P., and Minner, D. 1981. Evaluation of fungicides for preventive and curative control of Fusarium blight on Kentucky bluegrass, 1980. Fungic. Nematic. Tests 36:145.

20. Dernoeden, P. H., and Nash, A. S. 1982. Evaluation of fungicides for curative control of Fusarium blight on Kentucky bluegrass, 1981. Fungic. Nematic. Tests 37:151.

21. Dernoeden, P. H., and O'Neill, N. R. 1983. Occurrence of Gaeumannomyces patch disease in Maryland and growth and pathogenicity of the casual agent. Plant Dis. 67:528-532.

22. Dernoeden, P., O'Neill, N., and Murray, J. 1981. Evaluation of fungicides for control of Ophiobolus patch and residual effects against dollar spot, 1980. Fungic. Nematic. Tests 36:136.

23. Elliott, M. L. 1991. Determination of an etiological agent of bermudagrass decline. Phytopathology 81:1380-1384.

24. Elliott, M. L. 1992. Cultural and chemical control of bermudagrass decline caused by *Gaeumannomyces graminis* var. *graminis*. (Abstr.) Phytopathology 82:1123.

25. Elliott, M. L., and Freeman, T. E. 1991. Bermudagrass decline. Fact Sheet PP-31. University of Florida, Gainesville.

26. Elliott, M. L., and Landschoot, P. J. 1991. Fungi similar to *Gaeumannomyces* associated with root rot of turfgrasses in Florida. Plant Dis. 75:238-241.

27. Endo, R. M., and Colbaugh, P. F. 1974. Fusarium blight of Kentucky bluegrass in California. Pages 325-327 in: Proc. Int. Turfgrass Res. Conf., 2nd. E. C. Roberts, ed. American Society of Agronomy and Crop Science Society of America, Madison, WI.

28. Endo, R. M., Ohr, H. D., and Krausman, E. M. 1984. The cause of the spring dead spot disease (SDS) of *Cynodon dactylon* (L.) Pers. in California. (Abstr.) Phytopathology 74:812.

29. Endo, R. M., Ohr, H. D., and Krausman, E. M. 1985. *Leptosphaeria korrae*, a cause of the spring dead spot disease of bermudagrass in California. Plant Dis. 69:235-237.

30. Fulton, D. E., Cole H., Jr., and Nelson, P. E. 1974. Fusarium blight symptoms on seedling and mature Merion Kentucky bluegrass plants inoculated with

Fusarium roseum and *Fusarium tricinctum*. Phytopathology 64:354-357.

31. Goss, R. L., and Gould, C. J. 1967. Some interrelationships between fertility levels and Ophiobolus patch disease in turfgrasses. Agron. J. 59:149-151.

32. Gould, C. J., Goss, R. L., and Eglitis, M. 1961. Ophiobolus patch disease of turf in western Washington. Plant Dis. Rep. 45:296-297.

33. Hartman, J. R., Clinton, W., and Powell, A. J. 1989. Control of summer patch and necrotic ring spot of Kentucky bluegrass, 1988. Fungic. Nematic. Tests 44:248.

34. Jackson, N. 1958. Ophiobolus patch disease fungicide trial, 1958. J. Sports Turf Res. Inst. 9:459-461.

35. Jackson, N. 1979. More turf diseases; old dogs and new tricks. J. Sports Turf Res. Inst. 55:163-166.

36. Juska, F. V., and Murray, J. J. 1974. Performance of bermudagrass in the transition zone as affected by potassium and nitrogen. Pages 149-154 in: Proc. Int. Turfgrass Res. Conf., 2nd. E.C. Roberts, ed. American Society of Agronomy and Crop Science Society of America, Madison, WI.

37. Kackley, K. E., Grybauskas, A. P., Dernoeden, P. H., and Hill, R. L. 1990. Role of drought stress in the development of summer patch in field-inoculated Kentucky bluegrass. Phytopathology 80:655-658.

38. Kackley, K. E., Grybauskas, A. P., Hill, R. L., and Dernoeden, P. H. 1990. Influence of temperature-soil water status interactions on the development of summer patch in *Poa* spp. Phytopathology 80:650-655.

39. Kemp, M. L. 1991. The susceptibility of fine fescues to isolates of *Magnaporthe poae* and *Gaeumannomyces incrustans*. Ph.D. dissertation. Rutgers University, New Brunswick, NJ.

40. Kemp, M. L., Clarke, B. B., and Funk, C. R. 1990. The susceptibility of fine fescues to isolates of *Magnaporthe*

poae and *Gaeumannomyces incrustans.* (Abstr.)
Phytopathology 80:978.

41. Kemp, M. L., Landschoot, P. J., Clarke, B. B., and Funk, C. R. 1990. Response of fine fescues to field inoculation with summer patch. Page 176 in: Agronomy Abstracts. American Society of Agronomy, Madison, WI.

42. Kozelnicky, G. M. 1974. Updating 20 years of research: spring dead spot. USGA Green Section Rec. 12:12-15.

43. Landschoot, P. J., and Clarke, B. B. 1989. Evaluation of fungicides for control of summer patch on annual bluegrass, 1988. Fungic. Nematic. Tests 44:245.

44. Landschoot, P. J., and Hoyland, B. F. 1992. Effect of various nitrogen sources on necrotic ring spot development, 1990-1991. Biol. Cultural Tests 7:109.

45. Landschoot, P. J., and Hoyland, B. F. 1992. Resistance of Kentucky bluegrass cultivars and selections to necrotic ring spot, 1991. Biol. Cultural Tests 7:110.

46. Landschoot, P. J., and Jackson, N. 1987. The *Phialophora* state of a *Magnaporthe* sp. causes summer patch disease of *Poa pratensis* L. and *P. annua* L. (Abstr.) Phytopathology 77:119.

47. Lucas, L. T. 1980. Control of spring dead spot of bermudagrass with fungicides in North Carolina. Plant Dis. 64:868-870.

48. Lucas, L. T. 1980. Spring deadspot of bermudagrass. Pages 183-187 in: Advances in Turfgrass Pathology. P. O. Larsen and B. G. Joyner, eds. Harcourt Brace Jovanovich, Duluth, MN.

49. Lucas, L. T., and Newnam, M. R. 1989. Evaluation of fungicides for the control of spring dead spot of bermudagrass, 1988. Fungic. Nematic. Tests 44:244.

50. Lucas, L. T., and Newnam, M. R. 1990. Evaluation of fungicides for control of spring dead spot of bermudagrass, 1989. Fungic. Nematic. Tests 45:272.

51. McCarty, L. B., and Lucas, L. T. 1988. Identification and suppression of spring dead spot disease in

bermudagrass. Page 154 in: Agronomy Abstracts. American Society of Agronomy, Madison, WI.

52. McCarty, L. B., and Lucas, L. T. 1989. *Gaeumannomyces graminis* associated with spring dead spot of bermudagrass in the southeastern United States. Plant Dis. 73:659-661.

53. McCarty, L. B., Lucas, L. T., and DiPaola, J. M. 1992. Spring dead spot occurrence in bermudagrass following fungicide and nutrient applications. HortScience 27:1092-1093.

54. McCarty, L. B., Lucas, L. T., and Newnam, M. R. 1988. Evaluation of fungicides for control of spring dead spot in bermudagrass, 1987. Fungic. Nematic. Tests 43:265.

55. Nash, A. S. 1988. The influence of nitrogen source and rate, application timing and irrigation on the quality of Kentucky bluegrass and tall fescue. M.S. thesis. University of Maryland, College Park, MD.

56. Pair, J. C., Crowe, F. J., and Willis, W. G. 1986. Transmission of spring dead spot disease of bermudagrass by turf/soil cores. Plant Dis. 70:877-878.

57. Plumley, K. A., Clarke, B. B., Hillman, B. I., and Bunting, T. E. 1992. The effect of mowing height on the distribution of *Magnaporthe poae* in the soil profile and the development of a DNA probe for the detection of this pathogen. (Abstr.) Phytopathology 82:1160.

58. Plumley, K. A., Clarke, B. B., and Landschoot, P. J. 1991. Evaluation of fungicides for the control of summer patch in annual bluegrass. (Abstr.) Phytopathology 81:124.

59. Sanders, P. L., and Soika, M. D. 1990. Control of necrotic ring spot on Kentucky bluegrass, 1989. Fungic. Nematic. Tests 45:278.

60. Sanders, P. L., and Soika, M. D. 1991. Control of necrotic ring spot on Kentucky bluegrass, 1990. Fungic. Nematic. Tests 46:313.

61. Sarniguet, A., and Lucas, P. 1991. Evolution of bacterial populations related to decline of take-all patch on turfgrass. (Abstr.) Phytopathology 81:1202.

62. Schoevers, T. A. C. 1937. Some observations on turf-diseases in Holland. J. Board Greenskeep. Res. 5:23-26.

63. Smiley, R. W. 1980. Fusarium blight of Kentucky bluegrass: New Perspectives. Pages 155-175 in: Advances in Turfgrass Pathology. P. O. Larsen and B. G. Joyner, eds. Harcourt Brace Jovanovich, Duluth, MN.

64. Smiley, R. W. 1984. "Fusarium blight syndrome" re-described as a group of patch diseases caused by *Phialophora graminicola, Leptosphaeria korrae*, or related species. (Abstr.) Phytopathology 74:811.

65. Smiley, R. W. 1987. The etiologic dilemma concerning patch diseases of bluegrass turfs. Plant Dis. 71:774-781.

66. Smiley R. W., and Cook, R. J. 1973. Relationship between take-all of wheat and rhizosphere pH in soils fertilized with ammonium vs. nitrate-nitrogen. Phytopathology 63:882-890.

67. Smiley, R. W., and Craven Fowler, M. 1984. *Leptosphaeria korrae* and *Phialophora graminicola* associated with Fusarium blight syndrome of *Poa pratensis* in New York. Plant Dis. 68:440-442.

68. Smiley, R. W., and Craven Fowler, M. 1986. Necrotic ring spot and summer patch of Kentucky bluegrass sod monocultures, blends and mixtures, 1985. Biol. Cultural Tests 1:60.

69. Smiley, R. W., Dernoeden, P. H., and Clarke, B. B. 1992. Compendium of Turfgrass Diseases. 2nd ed. American Phytopathological Society, St. Paul, MN.

70. Smiley, R. W., and Fowler, M. C. 1985. Techniques for inducing summer patch symptoms on *Poa pratensis*. Plant Dis. 69:482-484.

71. Smiley, R. W., Fowler, M. C., and Kane, R. T. 1985. Temperature and osmotic potential effects on *Phialophora graminicola* and other fungi associated with patch diseases of *Poa pratensis*. Phytopathology 75:1160-1167.

72. Smith, A. M. 1965. *Ophiobolus herpotrichus*, a cause of spring dead spot in couch turf. Agric. Gaz. N. S. W. 76:753-758.

73. Smith, A. M. 1971. Control of spring dead spot of couch grass turf in New South Wales. J. Sports Turf Res. Inst. 47:60-65.

74. Smith, J. D. 1956. Fungi and turf diseases. (6) Ophiobolus patch disease. J. Sports Turf Res. Inst. 9:180-202.

75. Smith, J. D., Jackson, N., and Woolhouse, A. R. 1989. Fungal Diseases of Amenity Turf Grasses. E. & F. N. Spon, London.

76. Thompson, D. C., and Clarke, B. B. 1991. In vitro sensitivity of *Magnaporthe poae* and other root and crown-infecting fungi causing patch diseases of turfgrass to selected fungicides. (Abstr.) Phytopathology 81:1176.

77. Thompson, D. C., and Clarke, B. B. 1992. Evaluation of bacteria for biological control of summer patch of Kentucky bluegrass caused by *Magnaporthe poae*. (Abstr.) Phytopathology 82:1123.

78. Thompson, D. C., Heckman, J. R., and Clarke, B. B. 1992. Evaluation of nitrogen form and the rate of nitrogen and chloride application for the control of summer patch on 'Fylking' Kentucky bluegrass. (Abstr.) Phytopathology 82:248.

79. Tisserat, N. A., Pair, J., and Nus, A. 1988. Evaluation of fungicides for control of spring dead spot of bermudagrass, 1987. Fungic. Nematic. Tests 43:266.

80. Tisserat, N., Pair, J., and Nus, A. 1988. *Ophiosphaerella herpotricha* associated with spring dead spot of bermudagrass in Kansas. (Abstr.) Phytopathology 78:1613.

81. Tisserat, N. A., Pair, J., and Nus, A. 1989. Evaluation of fungicides for control of spring dead spot of bermudagrass, 1988. Fungic. Nematic. Tests 44:245.

82. Tisserat, N. A., Pair, J. C., and Nus, A. 1989. *Ophiosphaerella herpotricha*, a cause of spring dead spot of bermudagrass in Kansas. Plant Dis. 73:933-937.

83. Tisserat, N. A., Pair, J., and Nus, A. 1991. Evaluation of fungicides for control of spring dead spot of bermudagrass, 1990. Fungic. Nematic. Tests 46:306.

84. Turgeon, A. J. 1991. Turfgrass Management. 3rd ed. Prentice-Hall, Englewood Cliffs, NJ.

85. Turgeon, A. J., and Meyer, W.A. 1974. Effects of mowing height and fertilization level on disease incidence in five Kentucky bluegrasses. Plant Dis. Rep. 58:514-516.

86. Wadsworth, D. F., Houston, B. R., and Peterson, L. J. 1968. *Helminthosporium spiciferum*, a pathogen associated with spring dead spot of bermuda grass. Phytopathology 58:1658-1660.

87. Wadsworth, D. F., and Young, H. C., Jr. 1960. Spring dead spot of bermudagrass. Plant Dis. Rep. 44:516-518.

88. Walker, J., and Smith, A. M. 1972. *Leptosphaeria narmari* and *L. korrae* spp. nov., two long-spored pathogens of grasses in Australia. Trans. Br. Mycol. Soc. 58:459-466.

89. Wong, P. T. W., and Siviour, T. R. 1979. Control of Ophiobolus patch in *Agrostis* turf using avirulent fungi and take-all suppressive soils in pot experiments. Ann. Appl. Biol. 92:191-197.

90. Wong, P. T. W., and Worrad, D. J. 1989. Preventative control of take-all patch of bentgrass turf using triazole fungicides and *Gaeumannomyces graminis* var. *graminis* following soil fumigation. Plant Prot. Q. 4:70-72.

91. Worf, G. L., Brown, K. J., and Kachadoorian, R. V. 1983. Survey of "necrotic ring spot" disease in Wisconsin lawns. (Abstr.) Phytopathology 73:839.

92. Worf, G. L., Stewart, J. S., and Avenius, R. C. 1986. Necrotic ring spot disease of turfgrass in Wisconsin. Plant Dis. 70:453-458.

93.   Young, H. C., Jr., Sturgeon, R. V., Jr., and Huffine, W. W. 1973. Soil type associated with spring dead spot of bermudagrass. (Abstr.) Phytopathology 63:450.